The Dynamics of Discrete Populations and Series of Events

Graduate Student Series in Physics

Series Editor:

Professor Derek Raine

Senior Lecturer, Department of Physics and Astronomy, University of Leicester

The Dynamics of Discrete Populations and Series of Events
Keith Iain Hopcraft, Eric Jakeman, and Kevin D. Ridley

Introduction to Surface and Superlattice Excitations, Second Edition
Michael G. Cottam and D.R. Tilley

Cosmology in Gauge Field Theory and String Theory
D. Bailin and Alexander Love

Introduction to Nuclear Reactions
C.A. Bertulani and P. Danielewicz

Gauge Theories in Particle Physics, Volume II: QCD and the Electroweak Theory, Third Edition
I.J.R. Aitchison and A.J.G. Hey

Gauge Theories in Particle Physics: Volume I: From Relativistic Quantum Mechanics to QED, Third Edition
I.J.R. Aitchison and A.J.G. Hey

Problems on Statistical Mechanics
D.A.R Dalvit, J. Frastai, and Ian Lawrie

Symmetries in Quantum Mechanics: From Angular Momentum to Supersymmetry (PBK)
M. Chaichian and R. Hagedorn

Supersymmetric Gauge Field Theory and String Theory
D. Bailin and Alexander Love

Introduction to Gauge Field Theory Revised Edition
D. Bailin and Alexander Love

Superfluidity and Superconductivity
D.R. Tilley and J. Tilley

The Dynamics of Discrete Populations and Series of Events

Keith Iain Hopcraft
University of Nottingham, UK
Eric Jakeman
University of Nottingham, UK
Kevin D. Ridley
Consultant Optical Physicist, UK

CRC Press
Taylor & Francis Group
Boca Raton London New York

CRC Press is an imprint of the
Taylor & Francis Group, an **informa** business

CRC Press
Taylor & Francis Group
6000 Broken Sound Parkway NW, Suite 300
Boca Raton, FL 33487-2742

First issued in paperback 2019

ISBN-13: 978-1-4200-6067-6 (hbk)
ISBN-13: 978-0-367-37895-0 (pbk)

Library of Congress Cataloging-in-Publication Data

Hopcraft, K. I.
The Dynamics of discrete populations and series of events / Keith Iain Hopcraft, Eric Jakeman, Kevin D. Ridley.
pages cm. -- (Graduate student series in physics)
"A CRC title."
Includes bibliographical references and index.
ISBN 978-1-4200-6067-6 (hardcover : alk. paper)
1. Markov processes. 2. Mathematical statistics. 3. Variables (Mathematics) I. Jakeman, E. II. Ridley, K. D. (Kevin Dennis) III. Title.

QA274.7.H66 2014
519.2'33--dc23 2013045990

Visit the Taylor & Francis Web site at
http://www.taylorandfrancis.com

and the CRC Press Web site at
http://www.crcpress.com

Contents

Preface

This book is largely concerned with mathematical models that can be used to describe time-varying populations and series of events. These models are simple and generic, and the emphasis is on their general properties rather than their applicability to any particular real-world systems, although examples of the latter can be found in some of the further reading material listed at the end of each chapter. Although the models are relatively simple and analytically tractable, they provide a basic understanding of the effect of the competing processes governing the changing number of individuals present in a population and therefore a guide to the development of more complex models that can be accessible only through numerical modelling.

The book was originally stimulated by the recent development of models that can be used to characterise the evolution of populations with fluctuations governed by long-tailed probability distributions such as are commonly observed in so-called complex systems. It was clear to the authors that a transparent description of such models would be greatly facilitated by tracing the way that they had emerged from the more familiar population processes described in the existing literature. The main part of this book is therefore devoted to the logical development of the theory of first-order discrete Markov population processes starting from a few basic assumptions. Note the term *discrete*, here; it is also possible to characterise populations by approximating them with continuous variables, and this is a method that has been widely used. However, this book is concerned solely with the discrete approach. More information about these two contrasting methods can be found in Chapter 1.

The book also covers work inspired by the authors' background in quantum optics. This led them to extend the more familiar models to situations where the populations have novel properties, including sub-Poisson statistics and odd–even effects that have no continuum counterpart. Perhaps more importantly, it prompted them to investigate processes by which the populations may be monitored and to calculate how the measured and intrinsic statistics differ. This leads naturally to the generation and characterisation of time series that record, for example, when individuals leave a population.

A substantial review of the history of the subject and of the contents of each chapter is given in the introductory Chapter 1, but the book does not aim to give a comprehensive coverage of what is a very large area of science with a long and distinguished history and with a wide range of applications. Rather, it is intended to provide some basic mathematical tools, physical insight, and some novel ideas that will assist students to develop their own stochastic modelling capabilities. To aid this, Chapter 11 describes methods for numerical simulation of the models developed in the rest of the book, with examples given in the mathematical programming language *Mathematica*®. These examples are, however, kept sufficiently simple that even those with no in-depth knowledge of *Mathematica* should be able to re-write them in a programming language of their choice.

The authors are particularly indebted to Terry Shepherd, and to Eric Renshaw who has authored two comprehensive volumes of related work. They are also indebted, of course, to a number of research students for their work in this area, including S. Phayre, J.O. Matthews, O.E. French, J.M. Smith, and W.H. Lee.

1

Introduction

This book has a dual purpose. The first is to provide an introductory text that collects various discrete stochastic population models. This will furnish the reader with a systematic methodology for the formulation of the models, and it will explore their dynamical properties and the ways of characterising their behaviours in terms of customary measurements that can be made upon them. The second purpose is to then use these models as a tool for generating a series of events, or *point processes,* that have distinct properties according to which population model is used as a motor. The two qualifiers, *discrete* and *stochastic,* simultaneously provide the subject with a richness of phenomenology and technical challenges for formulating, describing, and extracting those behaviours. This is because the population can only change by an integer amount, and those changes are triggered to occur at times not governed by the invariable ticks of a clock. These strictures are absent when adopting a more straightforward *continuous* and *deterministic* approach, whereby the population is described as a continuous density and time marches uniformly onward at a regular pace. This distinction between the two approaches also delineates between the source, character and strength of fluctuations. In the stochastic formulation the fluctuations are an *intrinsic* property of the population itself. The mechanisms causing the changes in population size are essentially non-perturbative in nature because they change the state of the population by values of finite size, and this is true even if the mathematical formulation of a particular mechanism is linear. In the deterministic approach, intrinsic fluctuations can only arise through non-linearity. Both approaches can be affected by the presence of extrinsic noise, but again this is essentially non-perturbative in the stochastic formulation.

The dynamics of populations has a long and varied history, both in terms of the subjects' development per se and for the stimulus it has provided to other seemingly unconnected areas. Indeed, the subject has a pervasiveness that can cross between scientific boundaries and even beyond the ambit of the physical sciences by virtue of its utility. It was the cleric and scholar Robert Malthus who, in his *Essay on the Principle of Population* in 1798 [1], considered the brakes to unfettered linear exponential population growth. His conclusion that 'the power of population is indefinitely greater than the power in the earth to produce subsistence in man' resonates down the ages in spirit if not through the predictive accuracy of his analysis. This essay stimulated, in 1801, the establishment of the first decennial national census in the United Kingdom in order to ascertain how much corn was required to feed the nation (and, serendipitously, to quantify the number of able-bodied men available to fight in the Napoleonic wars). The essay provided one of many stimuli to Charles Darwin's development of the theory of natural selection. Darwin recognised that the effects of competition for resources by species as described by Malthus provided a mechanism for their diversification, whereby 'favourable variations would tend to be preserved, and unfavourable ones to be destroyed. The results of this would be the formation of a new species. Here, then I had at last got a theory by which to work' [2].

Pierre-Francois Vehulst sought to model and thereby quantify the limitations to exponential growth to which Malthus alluded [3]. This introduced to the equation describing the continuous and deterministic evolution of population size the notion of a *carrying capacity* that embodies an ecosystem's ability to sustain such a population. The key ingredient of this 'logistic' equation is its non-linearity, which predicted that the population's size saturated eventually and established the timescale on which this occurred. It is not so much this result but a technical reinterpretation of the logistic equation itself, over a century after Vehulst, which is of interest and profound consequence. This reinterpretation of the governing equation recognised that there is a season in which an animal or plant species breeds or grows. To encapsulate this observation, time was no longer treated continuously but as a discrete variable, although still deterministically. In this way the logistic differential equation is transformed to a difference equation or iterative mapping. The mapping contains a parameter, related to the carrying capacity, whose value leads to very different classes of behaviour. These range from solutions similar to those obtained from the continuous equation, to periodic bi-stability where the population oscillates between two values, period-doubling and aperiodic behaviour sensitive to the initial conditions. The logistic mapping therefore provides a paradigm for deterministic chaos, as described in the review article by Robert May [4].

Descriptions of the events that are associated with various population models are also ubiquitous because of their usefulness. The model with which all others are compared is the Poisson process. The distribution for the number of these birth events occurring in an interval of time was introduced by Siméon-Denis Poisson but with totally different context and application, namely to the deliberations of juries in criminal and civil legal trials [5]. The Poisson process' reach is very wide because of its elemental and simple nature. The Poisson distribution is an example belonging to the discrete stable class of distributions. This means that the sum of Poisson distributed random variables is itself a Poisson distributed random variable. This makes the Poisson the discrete analogue of the continuous Gaussian random variable, which also possesses the stability property. The widening of the discrete stable class through the agency of population models, and the events associated with them, forms the subject matter of Chapter 7.

The intended readership of this book is the advanced undergraduate or postgraduate student, but we also have in mind the experienced researcher who may be changing fields, or who needs to know about a particular class of population model and to access key information for its characterisation. The mathematical machinery required to understand the development will have been encountered by most chemistry, mathematics, physics and theoretical biology students. Moreover, we have adhered to pursuing the development using a limited bag of tools, even in those instances when application of another method can produce a result with greater economy of effort. If a particular model admits an analytical solution, then that solution can usually be obtained by more than one means. The text is concerned with exploring the stochastic formulation of population models and the qualitative differences between them rather than the methods available for obtaining their solution. The following chapters include a brief summary, some problems that explore both the basics and some more interesting aspects of the development, and a 'Further Reading' list. A manual with solutions to the problems is also available. The bibliography is not meant to be exhaustive but rather a first port of call for the reader to explore applications, subtleties, and further techniques.

Chapter 2 provides a primer for the probabilistic and statistical tools that will be used throughout. It introduces the discrete distributions that will feature in subsequent chapters and also their alternative representation in terms of a generating function. The technique of using generating functions to represent the population dynamics may, at first encounter, appear as an unwelcome abstraction and diversion. But their use enables the equations to be solved in a systematic fashion, and once obtained, the probabilities and moments that correspond to observable quantities can be derived from the generating function by elementary means.

Chapter 3 commences by establishing the Markov property, which is a further preliminary probabilistic foundation that will inform the development of most of the population models we go on to describe. This property pertains to how the future evolution of a system is affected by its immediate or more distant past. The chapter continues by exploring the three important elements of births, deaths and immigration as a cause for population change, and shows, with the assistance of the Markov property, how these separate processes are represented mathematically in the equation describing the stochastic evolution. Rather than solve for the complete dynamics, the evolution of measurables, such as the mean size and the correlation in size between one instant and the next, is obtained. The reason for this is that these microscopic causes for change in a populations' size are easy to intuit in terms of their averaged manifestation; births and deaths occurring in a population cause it to either grow or diminish in size unless the birth and death rates are identical, and so this contains no particular surprises. The stochastic treatment of the birth process was originally treated by Yule [6], in the context of mathematical biology, and independently by Furry [7] with regard to transmutation of elements by radioactive decay. The combination of deaths with immigration leads to the important concept of a stationary or equilibrium state for the population. Indeed, this equilibrium forms something of a benchmark against which all others can be compared, for it is the ubiquitous Poisson distribution. The simple mechanisms of birth, death and immigration provide a useful ground in which to explore the effect of measurement. The idea of making a measurement on a fluctuating quantity is not as straightforward as one might expect, or indeed hope. Any measurement takes a finite time to perform, during which the population continues to change. If that change is small compared with the population's size at the commencement of the measurement, then it will approximate the instantaneous state of the population itself. If the converse is true, the measurement *process* will sample and aggregate the various intermediate states through which the population evolves during the measurement time. The type of measurement made can also affect the population itself. Counting the number of individuals leaving (say) an airport is fundamentally different from counting the number of photons leaving a cavity. The detection of a photon is synonymous with its destruction and it can no longer participate in the dynamics. The measurement process also furnishes a more abstract but utilitarian function for turning the fluctuations in population *size* into a series of events in *time* that mark when those changes occur. It is this connection between the primary process by which the population evolves in time, and the secondary process by which it is monitored, that forms an important and recurrent theme throughout the text.

Chapter 4 considers the simultaneous combination of births, deaths, and immigration into a process, and this leads to a broader class of equilibrium solution provided that the death rate exceeds the rate at which births occur. This problem has surfaced in numerous branches of science, with the three mechanisms for change being co-opted to represent genetic mutations, through the production of photons in a cavity to the fluctuating state of the sea's surface. The full machinery of the generating function method of solution is deployed here for the first time. This serves to illustrate, in a natural way, how the relative sizes of the rates affect the structure of the equilibrium solution, how these combine to form the timescale that governs the approach to this steady state and the intrinsic fluctuations in it. The monitoring of this process is explored in more detail with two scenarios being treated. The first of these is where the counted 'individuals' are removed and no longer participate in the evolution; the second corresponds to when they are replaced and so continue to contribute to the evolution. Although one might consider that such distinctions would lead to minor differences, the discrepancies are nevertheless significant and the underlying physical interpretation for their occurrence is explored. The equilibrium solution of the birth–death–immigration process is characterised by fluctuations whose relative sizes are greater than the benchmark Poisson process. The series of events that are generated by this process are characterised by their occurring in *clusters* or *bunches* and therefore have an intermittent quality.

There are processes whose fluctuations are of lesser size than those associated with the Poisson, and events whose occurrences show a greater regularity than the purely random, manifesting a propensity for *anti-bunching*. Such populations can be generated using the mechanisms already introduced but with the important modification that the population size is capped but dynamically coupled with the extant population size. A limiting mechanism of this kind is redolent of the carrying capacity of an environment, but crucially it has no continuum analogue and is therefore quite distinct from the logistic model. A *lower limit* can also be applied so that the population cannot fall below some base value, and even when the cap is removed, this process can exhibit sub-Poisson effects. Chapter 5 provides the details of these models that exhibit sub-Poisson traits whilst retaining similar mathematical properties to those processes already encountered. An altogether different process that also has no continuum equivalent arises from relaxing the condition that the population can only increase though singletons. If immigrants enter the population in pairs but die singly, then the fluctuations display different properties according to whether there are an even or odd number present. Moreover, the monitoring without replacement of such a population serves to amplify the odd–even parity effects. Although this process is of interest to a rather arcane amplifier effect in quantum optics, its formulation prompts an important generalisation of the death–immigration process to one possessing great flexibility for modelling arbitrary populations with a prescribed steady state.

Allowing the immigrants to enter not just singly or in pairs, but in addition as triplets, …, r-tuplets, with rates of immigration particular to r, results in a mathematical structure with a simple and appealing interpretation. The population may be thought as being coupled to a separate equilibrium population of potential immigrants; it provides a coupling to an environment. Crucially, the formulation, which is described in Chapter 6, admits an inverse problem to be performed, whereby a population with a desired equilibrium state can be constructed through tailoring the rates at which the multiple immigrants are introduced.

The utility of the death–multiple immigration model is demonstrated in Chapter 7 to generate the discrete analogue of the stable process. Continuous stable random variables were treated by Paul Lévy in 1937 [8] as a generalisation of the Gaussian random variable, which is ubiquitous in pure and applied science. Nevertheless, the generalisation resulted in distributions whose variance is infinite and for which there is no characteristic associated scale size. Consequently, the generalisation was regarded as mathematical pathology and of little practical importance. In 1963, Beniot Mandelbrot showed that the fluctuations in cotton prices were described by such distributions [9]. So too were the fluctuations in diverse systems whose correlation scale size become commensurate with that of the system, irrespective of the details of the dynamics governing the system itself. Such behaviours are observed in systems close to some change in its phase or state. Throughout the late 1980s onward, various manifestations of the 'sandpile' paradigm were developed with the aim of demonstrating manifestations of these phenomena without recourse to fine-tuning a systems' parameters in order to adopt particular outcomes, in other words to achieve *emergent* behaviours from a parsimonious and generic dynamics [10]. This was achieved using computational models, and these possessed the wild fluctuations and scale-free properties of the non-Gaussian stable distributions. Since these were computational results, they are necessarily discrete in nature. A discrete stable *process* can be generated using the death–multiple immigration population model. The particular problems associated with monitoring such a process are acute, for the detecting 'instrument' will have a finite range or bandwidth, whereas the population has fluctuations of arbitrary size. Saturation effects in sampling, which are therefore inevitable, must be accounted for. When monitored, the series of events so formed present fractal properties irrespective of the properties of the detector—the scale-free attributes being transferred from a cascade in increasing population size to an inverse cascade in the time between events.

The coupling of populations in the stochastic framework presents technical challenges that are absent in the continuous and deterministic picture. It is an important area with much scope for future work in diverse application areas. A celebrated continuous formulation was pioneered independently by Alfred Lotka and Vito Volterra and has become known as the *predator–prey equations* [11], through which the dynamics of an idealised ecosystem comprising the hunters and hunted can be explored. The developments of this paradigm to treat a more diverse, realistic, and necessarily more complex set of ecosystems is described by Robert May [12]. From the stochastic perspective, the incorporation of immigrations into a population, be they single or multiple, is essentially modelling the situation of the population being affected and adjusting to accommodate the influence of an external environment. This occurs without the population reacting back onto those stimuli. Chapter 8 examines how the multiple immigration picture can be adapted so that a population and its environment become fully coupled with one another, evolving self-consistently according to their internal and mutual dynamics. This approach admits treating increasingly complex situations, where the environment can be regarded as a heterogeneous ensemble of different populations that connected in a chain. But the interactions between populations can occur in a greater variety of ways. The situation where members of one population are exchanged and rebadged as members of another provides another mechanism that displays distinct characteristics.

Rather than a series of events being driven through the agency of a discrete population process, it is possible for the events to be caused by some underlying fluctuation in a continuous quantity. Useful ideas frequently emerge in unconnected application areas and at broadly similar times. David Cox formulated this idea and illustrated it with application to the periods when a weaving loom was operational, dormant or had broken down [13]. By contrast, Leonard Mandel independently arrived at the same formulation with regard to the intervals between detection of photons using photoelectric multiplier tubes [14]. Such a representation has become known as *doubly stochastic* or *compound* because the fluctuations in the continuous variable stimulate an associated discrete process, the two forming a transform pair. This mechanism is explored in Chapter 9 and furnishes a method of great power and flexibility for generating populations. Indeed, the method can be demonstrated using the discrete stable process detailed in Chapter 7 to deduce the continuous process that produces it. Not surprisingly (perhaps) this is a continuous stable *process,* and the underlying Markov assumption enables the entire joint statistical properties of stable distributed variables to be deduced within the doubly stochastic context.

Chapter 10 discusses another mechanism for generating a series of events defined by an underlying continuous random process exceeding a predefined threshold: the *level crossing problem.* This problem has a huge literature associated with it [15], which has continued to expand. The reason for the large volume and diversity of work is principally because of the problem's close association with diverse applications in communications theory and technology, instrumentation and signal processing, and its relevance to the size, frequency and duration of extremal events. The principal difference between this problem and all the others treated in the text stems from its non-Markovian nature; the future evolution of the continuous process is affected not just by its immediate past but also further back into its history. This complexity through the mechanism of 'memory' is inherited by the level crossings and properties associated with them. The mathematical manifestation of this occurs through the spectrum of the continuous process having a non-Lorenztian form, which leads to profound technical challenges but also considerable richness in phenomenology. The statistical properties of the events display all the attributes that particular models describe: bunching, anti-bunching, fractal and sub-fractal.

Despite the emphasis placed on analytical methods, computational simulation of populations and the properties derived from them provides an important complementary approach to understanding, model development and the treatment of real-world situations. Often it is the only way to make progress. Chapter 11 describes the techniques for the population models previously described, with examples of *Mathematica*® code for their implementation. The essentials of any stochastic population model are to determine when the next event occurs and its nature, both of these being conditioned on what happened at the preceding event. This procedure can be elaborated without difficulty to incorporate an arbitrary number of different mechanisms that influence the dynamics of the population. The algorithmic method usually carries Daniel Gillespie's name [16], following his treatment of chemical reactions, but it can be traced back further to a computational implementation made by David Kendall [17] of a birth–death population model that flowed from rigorous work in probability theory carried out by Andrey Kolmogorov, William Feller and Joseph Doob. The other numerical technique describes the simulation of a Gaussian random process with zero mean, unit variance and allowable but otherwise arbitrary autocorrelation function, hence a non-Markov situation. The higher-order statistics of a Gaussian process is completely defined through this single function. New continuous processes can be generated as functions of a Gaussian process, and so the ability to simulate them provides utility and flexibility for producing a range of continuous random processes.

The construction of models is the most basic activity in science. A 'good' model is one which captures the essentials of a particular phenomenon and which possesses a predictive capacity, both of these being achieved with the fewest of underlying assumptions. If those assumptions are transgressed or applied inappropriately, the model will fail to correspond to the reality. Hence, a model should not be confused with the actual phenomenon it purports to explain, rather it is a simulacrum of some aspect of its manifestation. This book contains a collection of models that have found a direct applicability in a range of application areas. But they also supply a more general way of thinking about the generation of events having different characters. We have aimed to show how these models can be fabricated in a hierarchical and systematic way from a few basic elements, and have arranged them according to the different phenomena they present. The reader can therefore select whether a given model has the appropriate character for describing their application of interest. If not, then a further aim is to provide for the reader the foundations and techniques to apply, adapt, and augment the models described here together with the ability and confidence to create new models for their own purposes.

References

1. Thomas Robert Malthus, *An Essay on the Principle of Population*, Oxford World's Classics, Oxford, 1798, 2008.
2. Nora Barlow, ed., *The Autobiography of Charles Darwin*, W.W. Norton & Co., New York, 1969.
3. Pierre-Francois Vehulst, 'Notice sur la loi que la population pursuit dans son accroissement,' *Correspondance mathematique et physique*, **10**, 113–121 (1838).
4. R.M. May, 'Simple mathematical models with very complicated dynamics,' *Nature*, **261**, 459–467 (1976).
5. Simeon-Denis Poisson, '*Recherches sur la probabilité des jugements en maitère criminelle et en matière civile,' Procédés des Règles Générales du Calcul Des Probabilités*, Bachelier, Imprimeur-Libraire pour les Mathématiques, Paris (1837).
6. G.U. Yule, 'A mathematical theory of evolution, based on the conclusions of Dr J.C. Willis,' *F.R.S. Philosophical Transactions of the Royal Society (London) Series B*, **213**, 21–87 (1924).
7. W.H. Furry, 'On fluctuation phenomena in the passage of high energy electrons through lead,' *The Physical Review*, **52**, 569–581 (1937).

8. P. Lévy, *Théorie de l'Addition des Variables Aléatories*, Gaultier-Villars, Paris, 1937.
9. B. Mandelbrot, 'The variation of certain speculative prices,' *The Journal of Business*, **36**, 394–419 (1963).
10. P. Bak, C. Tang, K. Weisenfeld, 'Self-organised criticality: An explanation of $1/f$ noise,' *Physical Review Letters*, **59**, 381–384 (1987).
11. A.J. Lotka, 'Contribution to the theory of periodic reaction,' *The Journal of Physical Chemistry*, **14**, 271–274, (1910). V. Volterra, 'Variations and fluctuations of the number of individuals in animal species living together,' in *Animal Ecology: With Special Reference to Insects*, R.N. Chapman, ed., McGraw-Hill, New York, 1931.
12. R.M. May, *Stability and Complexity in Model Ecosystems*, Princeton Landmarks in Biology Edition, Princeton University Press, Princeton, 2001.
13. D.R. Cox, 'Statistical methods connected with series of events,' *Journal of the Royal Statistical Society, Series B*, **17**, 129–164 (1955).
14. L. Mandel, 'Fluctuations of photon beams: The distribution of the photo-electrons,' *Proceedings of the Physical Society (London)*, **74**, 233–243 (1959).
15. I. Blake, W. Lindsay, 'Level-crossing problems for random processes,' *IEEE Transactions on Information Theory*, **19**, 295–315 (1973).
16. D.T. Gillespie, 'Exact stochastic simulation of coupled chemical reactions,' *The Journal of Physical Chemistry*, **81**, 2340–2361 (1977).
17. D.G. Kendall, 'An artificial realisation of a simple "birth-and-death" process,' *Journal of the Royal Statistical Society, Series B*, **12**, 116–119 (1950).

2

Statistical Preliminaries

2.1 Introduction

The purpose of this chapter is to provide a brief introduction to the statistical quantities and notation that will be used in the remainder of the book. The treatment is oriented to the case of *discrete* variables and is not intended to be exhaustive. The reader is referred to books listed under 'Further Reading' at the end of the chapter for more comprehensive treatments of the basics and particularly for situations in which the quantities of interest are not quantised.

After defining the basic concept of probability in the context of discrete *variables*, we shall describe some commonly encountered discrete probability distributions and characteristic measures by which their properties can be recognised and assessed. The *generating functions* that can often be used to simplify calculation will be introduced, and it will be shown how these can be used to recover both the distributions themselves and other useful properties. Discrete *processes* will be introduced through the concept of a population of individuals that is evolving with time. The number in the population at any given instant may then be related to that at previous times, giving rise to the notion of *correlation*. Another kind of discrete process is the *series of events* and in later chapters we shall show how simple mathematical models for the evolution of the number of individuals in a population can also be used to generate time sequences of this type. In the present chapter, two related measures of such a process will be discussed: the probability distribution of the number of events that occur in a given time interval and the statistics of the separation of the points in time at which the events occur.

2.2 Probability Distributions

Consider a large collection or *ensemble* of similar populations. We can work out the fraction of the populations containing exactly N individuals. This represents the *probability* $P_N \equiv P(N)$ of finding a population of N individuals in such an ensemble and we can plot this as a function of N as illustrated in

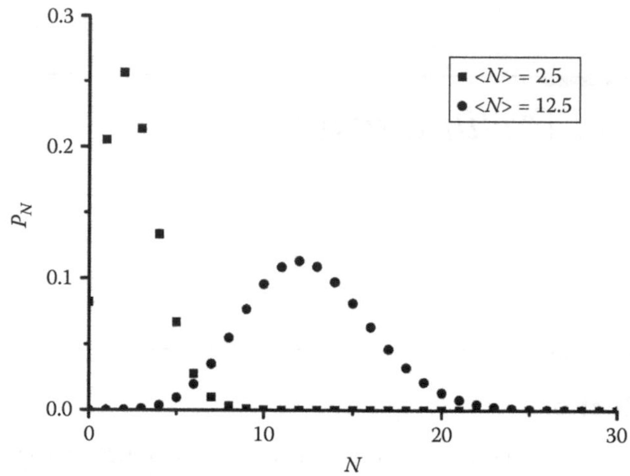

FIGURE 2.1
Poisson distribution for two different values of the mean.

Figure 2.1. This may well have the typical bell shape familiar from the theory of continuous random variables, but in the present case the abscissa measure will be discrete and the graph will be a sequence of discrete points. Since P_N is based on the *fraction* of the total number of populations containing N individuals, it will sum to unity:

$$\sum_{N=0}^{\infty} P_N = 1. \tag{2.1}$$

This *normalisation* requirement will often provide a useful check on the validity of results obtained in calculations. We note that it is reasonable to suppose that the probability of finding increasingly large numbers in the populations will reduce, so that the tail of the distribution will ultimately approach zero. However, no such *a priori* argument can be applied for the form of the graph for small values of the population number.

The distribution plotted in Figure 2.1 is commonly encountered in the theory of random numbers, and it is called the Poisson distribution:

$$P_N = \frac{\bar{N}^N}{N!}\exp(-\bar{N}). \tag{2.2}$$

This distribution has just a single free parameter, which is the *average* or *mean* number of individuals. For any distribution P_N, the average is defined as

$$\bar{N} = \sum_{N=0}^{\infty} N P_N = \langle N \rangle. \tag{2.3}$$

This also shows the two notations used in this book: the overbar and the angle brackets. The former is more compact, but the latter gives more clarity in some situations. These are used to denote the *ensemble average* of any quantity of interest, that is, for any function f of the population number

$$\overline{f}(N) = \sum_{N=0}^{\infty} f(N)P_N = \langle f(N) \rangle. \tag{2.4}$$

The Poisson distribution is found to provide a good model for a wide range of phenomena in science and engineering that can be characterised by random discrete variables. It is the large number limit of the well-known *binomial* distribution that describes the outcome of a number of trials involving two mutually exclusive events. If ξ is the chance of a particular result in a single trial then the probability of obtaining this result on N occasions in a total of M independent trials is given by

$$\begin{aligned} P_N &= {}^M C_N \xi^N (1-\xi)^{M-N} \qquad N \le M \\ &= 0 \qquad\qquad\qquad\qquad\quad N > M. \end{aligned} \tag{2.5}$$

Here the binomial coefficient is defined in the conventional way by the following ratio of factorials:

$${}^M C_N \equiv \begin{pmatrix} M \\ N \end{pmatrix} = \frac{M!}{N!(M-N)!}. \tag{2.6}$$

The most familiar scenario where Equation (2.5) would apply is in the tossing of a coin and in this case we would normally assume that $\xi = 1/2$, that is, in a single trial heads and tails would be equally likely. In the next section it is shown that the distribution (2.5) has a mean value $\overline{N} = M\xi$ and reduces to Equation (2.2) for large $M >> N$ when ξ is scaled to keep this mean value finite.

The binomial model Equation (2.5) can be generalised by including more than two mutually exclusive events. This leads to the *multinomial* distribution rather than the binomial distribution. However, in the present context a more important model that we shall frequently encounter in later chapters is provided by the *negative binomial* distribution,

$$P_N = \begin{pmatrix} N+\alpha-1 \\ N \end{pmatrix} \frac{(\overline{N}/\alpha)^N}{(1+\overline{N}/\alpha)^{N+\alpha}}. \tag{2.7}$$

In this formula the negative binomial coefficient is defined by analogy with Equation (2.6):

$$\begin{pmatrix} N+\alpha-1 \\ N \end{pmatrix} = \frac{\Gamma(N+\alpha)}{N!\Gamma(\alpha)}, \tag{2.8}$$

where the gamma or factorial function is defined in the usual way by

$$\Gamma(z) = \int_0^\infty dt\, t^{z-1} \exp(-t). \tag{2.9}$$

Unlike the Poisson distribution (2.2), the negative binomial distribution (2.7) contains a parameter or *index* α (which is always greater than zero) in addition to the average number of individuals. In fact Equation (2.7) defines a family or *class* of distributions that approach Poisson when α is very large. This can be seen from Figure 2.2 where a number of members of the class characterised by different values of α but the same mean value are plotted for comparison with the Poisson distribution. It is evident that, unlike the Poisson distribution, some members of the negative binomial class are monotonically decreasing for any value of the mean, with zero being the most probable number of individuals in the population. Moreover, the tails of the distributions decrease more slowly than that of the Poisson distribution. The most commonly encountered member of the negative binomial class is the case α = 1 when Equation (2.7) reduces to the *thermal, geometric* or *Bose-Einstein* distribution

$$P_N = \frac{\bar{N}^N}{\left(1+\bar{N}\right)^{N+1}}. \tag{2.10}$$

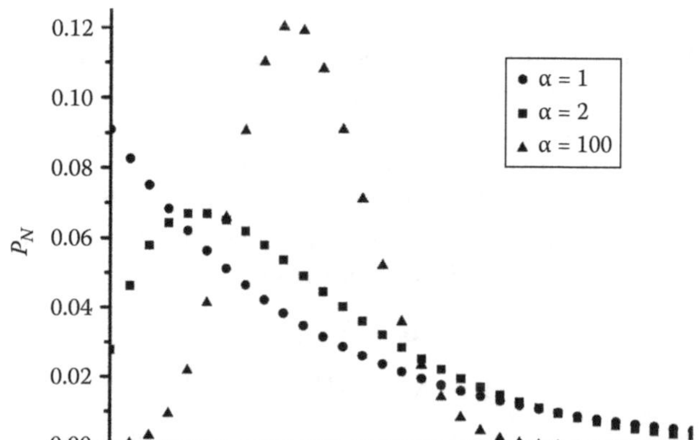

FIGURE 2.2
Negative binomial distribution for a mean value of 10 and three different values of the parameter α.

This distribution occurs widely in the physical sciences, particularly in quantum optics where it describes the distribution of photons in thermal light.

A less commonly encountered class of distributions that are important in the theory of photoelectron counting statistics are based on the Laguerre polynomials:

$$P_N = \frac{\alpha^\alpha N_1^N \exp\left[-\alpha N_2/(\alpha+N_1)\right]}{(\alpha+N_1)^{N+\alpha}} L_N^{\alpha-1}\left(\frac{-\alpha^2 N_2}{N_1(\alpha+N_1)}\right). \tag{2.11}$$

When $N_2 = 0$ this reduces to the negative binomial class (2.7) since

$$L_N^{\alpha-1}(0) = \begin{pmatrix} N+\alpha-1 \\ N \end{pmatrix}. \tag{2.12}$$

Although it is by no means obvious by inspection, we shall show in the next section that Equation (2.11) reduces to the Poisson distribution (2.2) when N_1 is zero. Thus the Laguerre distribution can be used to interpolate between the negative binomial and Poisson models.

It is not difficult to show that both the Poisson distribution and the negative binomial distribution decrease rapidly with N for large values of N. For large mean values the Poisson distribution has a Gaussian shape whilst the negative binomial distribution falls off exponentially. In the context of complex systems, however, discrete models that decrease more slowly than this are required. In particular, distributions that fall off like a power of the number of individuals rather than in an exponential fashion prove to be useful. A simple analytical example of such a distribution is provided by the expansion for the generalised Riemann zeta function

$$(m+\nu,\alpha) = \sum_{n=0}^{\infty} \frac{1}{(n+\alpha)^{m+\nu}}. \tag{2.13}$$

Here $0 < \nu < 1$ and m is a positive integer. The definition (2.13) allows us to postulate the distribution

$$P_N = \frac{1}{(m+\nu,\alpha)} \frac{1}{(N+\alpha)^{m+\nu}}. \tag{2.14}$$

The parameter α is a characteristic scale of this distribution beyond which it decays like an inverse power law and in the limit $N >> \alpha$ the distribution is said to be *scale-free*. This may be contrasted with the thermal distribution (2.10), for example, whose behaviour at large N is proportional to $\exp(-N/\bar{N})$ and is therefore governed by the mean value of the distribution. Indeed, though Equation (2.14) is perfectly well defined when $m = 1$, for example, the mean value of the distribution in this particular case is infinite!

Finally, in this section we note that by making the simple shift $N \to N - L$ in the preceding distributions we obtain new models $P_N^{(L)}$ in which the probability of finding $N < L$ is zero. These find application in situations where there is a fixed lower bound on the number of individuals in a population and may be expressed as the convolution

$$P_N^{(L)} = \sum_{R=0}^{N} \delta_{RL} P_{N-R}, \tag{2.15}$$

where δ_{RL} is the Kronecker delta function, equal to unity when the suffixes are the same but zero otherwise.

2.3 Moment-Generating Functions

In the last section some simple examples of discrete probability distributions were described that were typically characterised by a limited number of parameters. In practice these parameters can often be determined most easily from measurements of the low order *moments* of the distributions, if these exist, rather than by parameter fits to the distributions themselves. The most frequently quoted quantity is the *variance* of the distribution, $\operatorname{var} N = \langle N^2 \rangle - \langle N \rangle^2$, which provides a measure of the width of the distribution. For example, it is not difficult to show for the negative binomial distribution (2.7) that

$$\frac{\langle N(N-1) \rangle}{\langle N \rangle^2} = 1 + \frac{1}{\alpha}. \tag{2.16}$$

Evidently a measurement of the quantity on the left of this equation would provide a rather straightforward estimate of the parameter α. This observation leads us to define a useful parameterisation of discrete distributions, namely, the *factorial moments:*

$$\langle N(N-1)(N-2)\ldots(N-R+1) \rangle = \sum_{N=R}^{\infty} N(N-1)(N-2)\ldots(N-R+1) P_N. \tag{2.17}$$

In practice the normalised forms of these functions are often more useful because these are usually independent of the mean:

$$N^{[R]} \equiv \frac{\langle N(N-1)(N-2)\ldots(N-R+1) \rangle}{\langle N \rangle^R}. \tag{2.18}$$

The left-hand side of result (2.16) is thus the *second* ($R = 2$) normalised factorial moment of the negative binomial distribution (2.7).

Although factorial moments can obviously be calculated directly provided the analytical form of the distribution in question is known, it is often easier to obtain them from a 'conjugate' mathematical representation known as the *moment-generating function*. This quantity plays a similar role for discrete distributions as the Fourier transform or *characteristic* function plays for the probability densities describing continuous variables. In this book we shall define the moment-generating function $Q(s)$ of the distribution P_N by the infinite sum

$$Q(s) = \sum_{N=0}^{\infty} P_N (1-s)^N = \left\langle (1-s)^N \right\rangle. \tag{2.19}$$

This definition is used in quantum optics and leads to slightly simpler formulae for commonly encountered distributions than the probability-generating function defined with $(1 - s)$ replaced by s. Note that the normalisation condition (2.1) becomes simply

$$Q(0) = 1. \tag{2.20}$$

It is not difficult to show by repeated differentiation that

$$P_N = \frac{1}{N!}\left(-\frac{d}{ds}\right)^N Q(s)\Big|_{s=1}, \tag{2.21}$$

$$N^{[R]} = \frac{1}{\bar{N}^R}\left(-\frac{d}{ds}\right)^R Q(s)\Big|_{s=0}. \tag{2.22}$$

The right-hand side of these two formulae are essentially the coefficients in Taylor expansions of the generating function about the points $s = 1$ and $s = 0$, respectively, and this often provides an alternative method of calculation to straight differentiation if the functional form of the generating function is known. To demonstrate the utility of generating functions, we will now calculate results for the statistical models described in the last section.

Substituting the distribution (2.2) into (2.19) and summing over N gives the generating function for the Poisson distribution:

$$Q(s) = \exp\left(-\bar{N}s\right). \tag{2.23}$$

From Equation (2.22) it is evident that

$$N^{[R]} = 1. \tag{2.24}$$

In the case of the binomial distribution (2.5), formula (2.19) is the binomial expansion of $(1 + x)^M$. Making the appropriate identification of x, we find the result

$$Q(s) = (1 - \xi s)^M. \tag{2.25}$$

According to Equation (2.22) the mean of the corresponding distribution is given by $\bar{N} = M\xi$ and if this is fixed by scaling ξ with M, then in the limit of large M Equation (2.25) reduces to Equation (2.23). Expanding the right-hand side of Equation (2.25) with respect to s or alternatively differentiating according to Equation 2.22 we find the normalised factorial moments

$$N^{[R]} = (1 - N^{-1})(1 - 2N^{-1})\ldots(1 - [R-1]N^{-1}). \tag{2.26}$$

The negative binomial distribution (2.7) has the generating function

$$Q(s) = \frac{1}{(1 + s\bar{N}/\alpha)^{\alpha}}. \tag{2.27}$$

Using the binomial theorem to expand the right-hand side in powers of s or using Equation (2.22) gives the normalised factorial moments of this distribution in the form

$$N^{[R]} = \frac{\Gamma(R+\alpha)}{\alpha^R \Gamma(\alpha)}. \tag{2.28}$$

Thus in the thermal case $\alpha = 1$ we obtain the simple result

$$N^{[R]} = R!. \tag{2.29}$$

The simplifying property often conferred by the generating function representation is demonstrated particularly well in the case of the Laguerre distribution (2.11) for which Q takes the form

$$Q(s) = \frac{1}{(1 + sN_1/\alpha)^{\alpha}} \exp\left(-\frac{sN_2}{1 + sN_1/\alpha}\right). \tag{2.30}$$

It is now easy to see that when $N_2 = 0$ this reduces to the negative binomial result (2.27) whilst when $N_1 = 0$ it reduces to the Poisson case (2.23). The normalised factorial moments of the Laguerre distribution can be obtained by using the expansion

$$\frac{1}{(1-z)^{\alpha}} \exp\left(\frac{-xz}{1-z}\right) = \sum_{N=0}^{\infty} L_N^{\alpha-1}(x) Z^N. \tag{2.31}$$

In the present case we identify $z = -sN_1/\alpha$ and $x = -\alpha N_2/N_1$. The multiple derivatives (2.22) introduce an extra factor of $R!$ leading to the normalised factorial moments

$$N^{[R]} = R!(1+\alpha N_2/N_1)^{-R} L_R^{\alpha-1}(-\alpha N_2/N_1). \tag{2.32}$$

In contrast to these simple examples, the generating function for the distribution (2.14) is the normalised form of a less familiar object, namely, the *Lerch transcendent,* defined as

$$\Phi((1-s), m+\nu, a) = \sum_{n=0}^{\infty} \frac{(1-s)^n}{(n+a)^{m+\nu}} \tag{2.33}$$

However, the power law behaviour of the tail of the Lerch distribution (2.14) is also a feature of other distributions, in particular the class of *stable distributions.* These will be considered briefly in the next section and in detail in Chapter 7.

We shall often encounter situations in which the discrete variable of interest is the sum of two other variables. If these variables are statistically independent from each other then the probability distribution of the sum variable is the convolution of the distributions characterising its components

$$P_N = \sum_{R=0}^{N} P_R^{(1)} P_{N-R}^{(2)}. \tag{2.34}$$

Defining generating functions in the usual way, it is not difficult to show that the convolution (2.34) is simply the *product* of the generating functions for the convolved distributions:

$$Q(s) = Q^{(1)}(s)Q^{(2)}(s). \tag{2.35}$$

If we take $P_N^{(1)} = \delta_{NL}$ in Equation (2.34) we obtain formula (2.15), and (2.34) will give the shifted version of the distribution corresponding to $Q^{(2)}(s)$. Thus from Equation (2.35) we find that the corresponding generating function is

$$Q(s) = (1-s)^L Q^{(2)}(s) \tag{2.36}$$

2.4 Discrete Processes

So far we have considered only the statistics of a single discrete variable. In practice we may have data on the behaviour of two or more different variables characterised by the *joint distribution* $P(N_1, N_2, \ldots, N_R)$. This describes the joint probability of finding the values $\{N_j\} \equiv N_1, N_2, \ldots N_R$ for R different discrete variables. Sometimes values of the variables will be related to each other or *correlated*. For example, in the case of two variables one may be large when the other is small. This would be the case, for example, if the product of the two variables was always equal to a constant. A particularly important property in this context, therefore, is *statistical independence*. A collection of R variables are statistically independent if

$$P(\{N_j\}) = \prod_{n=1}^{R} P^{(n)}(N_n). \tag{2.37}$$

Here the suffix (n) on $P^{(n)}(N_n)$ is used to label the distributions of the individual variables since these may be different. According to the definition (2.19), relation (2.37) implies that the generating function of the joint distribution is also a product over the generating functions for the distributions of the individual variables:

$$Q(s_1, s_2, \ldots s_R) =$$

$$= \sum_{N=0}^{\infty} P(N_1, N_2, \ldots N_R)(1-s_1)^{N_1}(1-s_2)^{N_2} \ldots (1-s_R)^{N_R} \tag{2.38}$$

$$= \prod_{n=1}^{R} Q^{(n)}(s_n).$$

A simple consequence of statistical independence that can be obtained from (2.37) or (2.38) by inspection is that

$$\langle N_1 N_2 \ldots N_R \rangle = \prod_{n=1}^{R} \langle N_n \rangle. \tag{2.39}$$

In particular this means that the *correlation coefficient* of any two of the variables will vanish, for example,

$$\langle N_1 N_2 \rangle - \langle N_1 \rangle \langle N_2 \rangle = 0. \tag{2.40}$$

However, the fact that two variables are uncorrelated does *not* guarantee their statistical independence since (2.38) implies that the correlations of all higher powers of the variables must also vanish.

Note that the generating function $Q_\Sigma(s)$ for the *sum* of statistically independent variables, $N_1 + N_2 + \cdots N_R$ is obtained by setting $s_n = s$ in (2.38) for all n, generalising the result (2.35). In the special case when the variables are also statistically identical so that $P^{(n)}(N_n) \equiv P(N_n)$ we then obtain

$$Q_\Sigma(s) = Q^R(s). \tag{2.41}$$

The distribution corresponding to Q is said to be *stable* if $Q_\Sigma(s) = Q(a_R s)$, where a_R is a constant, that is, if

$$Q^R(s) = Q(a_R s). \tag{2.42}$$

Thus the term *stable* refers to the property of distributions of sums of identical independent random variables that are essentially the same as those of the individual variables. Equation (2.42) is satisfied by functions of the form

$$Q(s) = \exp(-as^\nu) \qquad 0 < \nu \le 1. \tag{2.43}$$

The Poisson distribution (2.2) evidently corresponds to the case $\nu = 1$ (see Equation 2.23) and it is indeed well known that the sum of independent Poisson variables is also Poisson distributed. It will be shown in Chapter 7 that apart from this case all the members of the class defined by (2.43) possess power-law tails akin to the behaviour of the models (2.14) with $m = 0$.

Suppose now that the variables $\{N_j\}$ are the values taken at different times t_j by the number of individuals in a *single* evolving population of individuals. This is known as a *discrete process* $N(t)$. Clearly, such a process must consist of a series of jumps whereby N changes from one value to another. An example is given in Figure 2.3. If these jumps in the population have a random component we might attempt to characterise them by examining the evolution of many similar populations, calculating the ensemble statistics as before. Of course, we would have to ensure that the data was taken at a time that was statistically the same in the evolution of each population since the distribution $P(\{N_j\})$ will generally change with time. If only data from one

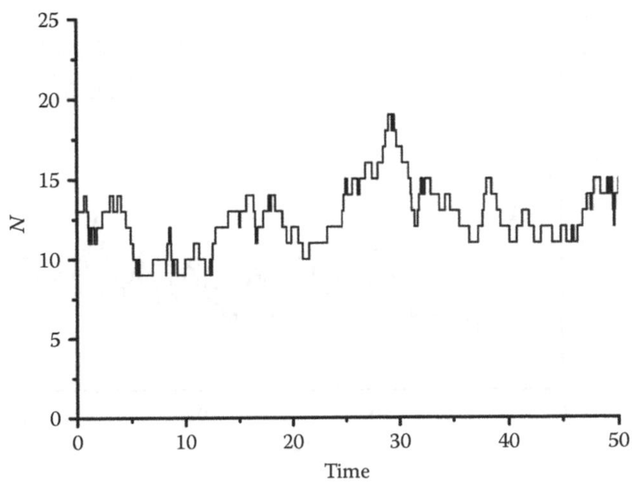

FIGURE 2.3
An example of a discrete process. This is the death–immigration process discussed in Chapter 3, Section 3.3. Here, $\mu = 0.1$ and $\nu = 1.0$.

population is available, however, the calculation of *time averages* provides an alternative approach:

$$\langle f(N)\rangle_T = \lim_{T\to\infty}\frac{1}{T}\int_{-T/2}^{T/2} dt\, f(N(t)). \tag{2.44}$$

When the probability distributions do not change with time, so that

$$P(\{N(t_j)\}) = P(\{N(t_j + t)\}) \tag{2.45}$$

for all values of t, the process governing the behaviour of $N(t)$ is said to be *stationary*. This does not mean that N is not changing with time but that the process has reached an equilibrium state in which the nature of the fluctuations is not changing with time, as illustrated in Figure 2.4. An important property implied by Equation (2.45) is that the correlation function or bilinear moment of the number of individuals is a function of the time-*difference* variable only:

$$\langle N(t_j)N(t_j+\tau)\rangle_T = \langle N(0)N(\tau)\rangle_T . \tag{2.46}$$

When time averages and ensemble averages are the same, so that for a function f

$$\langle f(N)\rangle = \sum_{N=0}^{\infty} f(N)P_N = \lim_{T\to\infty}\frac{1}{T}\int_{-T/2}^{T/2} f(N(t))\,dt, \tag{2.47}$$

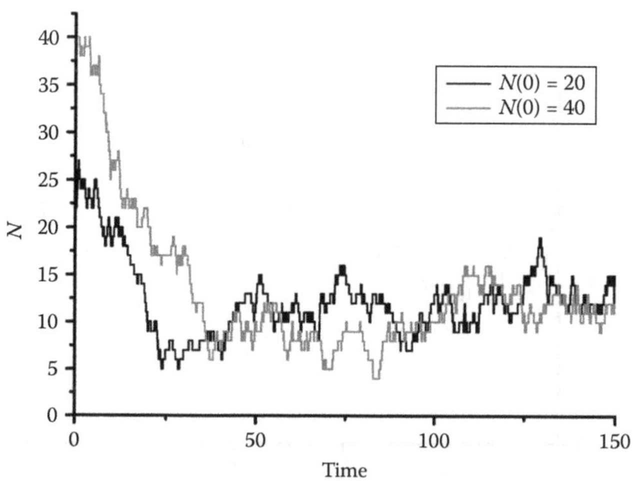

FIGURE 2.4
An death–immigration process starting from two different initial population numbers. After a transient interval, memory of the starting population is lost and both populations enter a stationary state. Here, $\mu = 0.1$ and $\nu = 1.0$.

the governing process is said to be *ergodic*. Ergodicity requires that all states of the system are accessible to each other through time evolution. Clearly, if this is not the case then some values may not be achieved in the time history of a single population although they may be found among the histories of a collection of populations subject to different initial conditions. In this book we shall be almost exclusively concerned with ergodic discrete processes.

2.5 Series of Events

In the last section we envisaged a particular kind of discrete stochastic process: the evolution of a population of individuals with time. This picture could be used to model how fluctuations in the number of people in a country or the number of bees in a hive, for example, develop as a consequence of the normal processes governing the life cycles of individuals in the respective populations. Likewise, it could be used to characterise the way that the number of photons in a laser cavity evolve in response to emission and absorption by the atomic system present, or the way that the number of molecules of a particular species change in a chemically reacting mixture. In principle it would be possible to statistically characterise real populations by simply counting the number of individuals at a succession of times. In practice this may be difficult, taking time during which the population changes, or it may interfere in some way with the evolution of the population so that the results of such a measurement would not give a true picture of the processes occurring.

In many situations of interest the way in which individuals physically leave the population also provides useful information about the process going on in the population itself. In this context the simplest characteristic that can be measured is the times at which individuals depart and this leads to the formulation of a new kind of process, namely, the *time series of events*. The properties of a series of identical events are completely defined by the times at which the events occur. Again, in practice this information may be difficult to obtain or record, and a simpler alternative is to count the number of events that occur in finite time intervals (*integration* or *counting times*) T. In order to shed light on the evolution of the original population it is necessary to know how this number changes with the 'running' time t during which the population is evolving. Thus we could measure the number $n(t_j;T)$ of counts in the interval T beginning at time t_j (Figure 2.5). The time instants $\{t_j\}$ could be chosen from the time history of the train of events in various ways, for example, randomly, periodically or logarithmically according to circumstances. In the case of a stationary process the

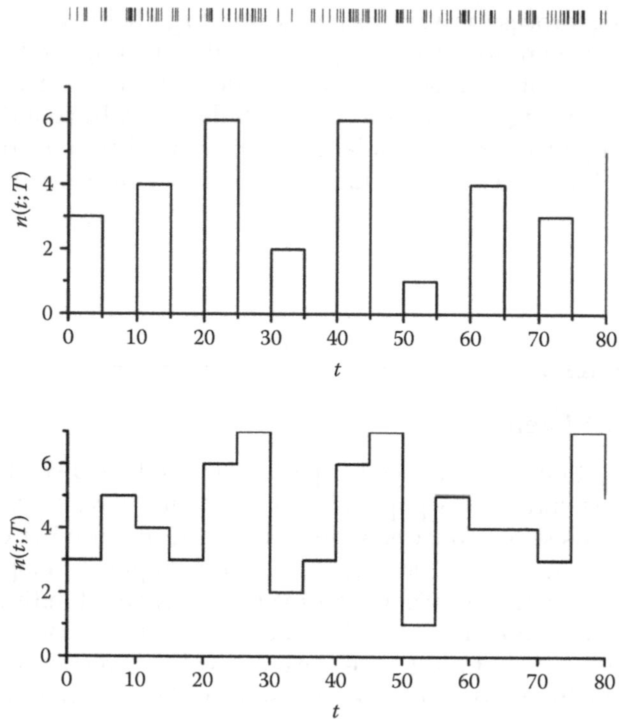

FIGURE 2.5
A series of events (top of picture) are counted in two different ways, both using an integration time of $T = 5$. First, through periodic sampling with period $2T$, and, second, contiguous sampling.

maximum amount of information is obtained in the continuous time limit when the quantity $n(t;T)$ constitutes a discrete process that is a continuous function of time but additionally parameterised by the integration time interval T.

The number $n(t;T)$ is characterised by the probability or *counting* distribution $p(n;T)$ of registering n events during an integration time (which will be independent of the running time in the case of a stationary process) and the corresponding generating function $q(s;T)$, moments and correlation properties of the kind we have already encountered in previous sections of this chapter. Note that we use lowercase letters to denote the analogous quantities. Thus, for example, we define the second normalised factorial moment of the event counting distribution by

$$n^{[2]}(T) = \frac{\langle n(n-1)\rangle}{\langle n\rangle^2}. \tag{2.48}$$

A statistical measure that provides a useful way of distinguishing the data from a purely random series of events is the *Fano factor*, defined as the variance of a distribution divided by its mean value, that is,

$$F = \frac{\langle n^2 \rangle - \langle n \rangle^2}{\langle n \rangle} \equiv 1 + \bar{n}\left(n^{[2]}(T) - 1\right). \tag{2.49}$$

This quantity is evidently unity when n is Poisson distributed. However, any deviation from unity of the second factorial moment is amplified by being multiplied by the mean. The Fano factor is a particularly sensitive measure in situations where variations in the event rate leading to non-Poisson behaviour occur on a timescale that is much shorter than the integration time T.

The way the statistics of $n(t;T)$ change with integration time can be related to the timing of the events and in particular to the statistics of the time to the first event and the time *between* events. To see this, consider a stationary process with the probability density of the time from an arbitrary time to that of the first event being $w_0(t)$ and designate $p(n;T)$ to be the probability of finding n events in the integration time T. Then by definition

$$\begin{aligned} \int_0^t dt' w_0(t') &= 1 - \Pr(\textit{first event has not been counted by time } t) \\ &= 1 - p(0;t). \end{aligned} \tag{2.50}$$

If we now define the generating function for $p(0;T)$ by analogy with Equation (2.19)

$$q(z;T) = \sum_{n=0}^{\infty} p(n;T)(1-z)^n = \left\langle (1-z)^n \right\rangle, \tag{2.51}$$

and take the derivative with respect to time of both sides of Equation (2.50) we find that

$$w_0(t) = -\frac{\partial q(1;t)}{\partial t}. \tag{2.52}$$

Suppose now that $w_1(t)$ is the probability density of the time between events. Then, if a time point is selected at random, the probability that it falls between two events separated by an interval of length between t and $t + \Delta t$ is $tw_1(t)\ \ t/\langle t \rangle$, where $\langle t \rangle$ is the mean interval duration. This is because the chance that the sampling point falls in such an interval is proportional to the total length of the interval that is proportional to $tw_1(t)$; the denominator being simply a normalisation constant. Now given that the sampling point falls in an interval of length t, it is equally likely to fall anywhere in that interval and hence the conditional probability density function of the time to the next event is uniform over $(0, t)$. Hence, the unconditional probability density of the time to the next event is just

$$w_0(t) = \int_t^\infty dt' \frac{1}{t'} \frac{t' w_1(t')}{\langle t \rangle}. \tag{2.53}$$

Taking the derivative with respect to t of both sides and using result (2.52) leads to

$$w_1(t) = \langle t \rangle \frac{\partial^2 q(1;t)}{\partial t^2}. \tag{2.54}$$

The simplest series of events would be one where they occur at a constant rate $r = \langle t \rangle^{-1}$ but at times that are statistically independent from one another. The number of events in a fixed time interval T will then be random and Poisson distributed. This may be shown by dividing the interval into small increments ΔT containing one or no events and applying the binomial result (2.25) and the result (2.41) for a sum of independent variables:

$$Q_\Sigma(s) = (1 - r \ \ Ts)^{T/\ T}. \tag{2.55}$$

Taking the limit of this result as $\Delta T \to 0$ gives

$$q(s;T) = \exp(-\bar{n}s). \tag{2.56}$$

According to (2.23) this is the generating function for a Poisson distribution with $\bar{n} = rT$ being the mean number of events in an integration time. It is now possible to calculate the probability densities (2.52) and (2.54):

$$w_0(t) = w_1(t) = r\exp(-rt). \tag{2.57}$$

Thus as might have been anticipated for a purely random series of events, both the time to the first event from an arbitrary time and the time between events have the same probability density. In later chapters we shall find that this is not the case for events that are not Poisson so that the Fano factor (2.49) differs from unity.

One further factor may often complicate measurements on a series of events: the *efficiency* of the counting process. Thus in many situations of interest, only a subset of events will be registered. If the counted events are selected at random, effectively by the toss of a coin, then the probability $p_m(n)$ of registering m out of n possible events will be given by the binomial coefficients that we encountered in Section 2.2:

$$p_m(n) = \binom{n}{m} \xi^m (1-\xi)^{n-m}. \tag{2.58}$$

Here, for a single event, ξ is the probability that it will be registered and $1 - \xi$ the probability that it will not be registered with $0 \le \xi \le 1$. ξ is therefore the efficiency of the measuring device or detector. This random or *Bernoulli* sampling scheme can fortunately be taken into account by a simple modification of the generating function for the ideal case when all the events are counted. To see this consider the modification of the ideal distribution $p(n,\xi = 1;T)$ caused by (2.58):

$$p(m,\xi;T) = \sum_{n=m}^{\infty} p_m(n) p(n,1;T). \tag{2.59}$$

The corresponding generating function is

$$q(s,\xi;T) = \sum_{m=0}^{\infty} (1-s)^m \sum_{n=m}^{\infty} p(n,1;T) \binom{n}{m} \xi^m (1-\xi)^{n-m}. \tag{2.60}$$

It is not difficult to rearrange the sums and demonstrate that

$$\begin{aligned} q(s,\xi;T) &= \sum_{n=0}^{\infty} \sum_{m=0}^{n} p(n,1;T) \binom{n}{m} (1-s)^m \xi^m (1-\xi)^{n-m} \\ &= \sum_{n=0}^{\infty} p(n,1;T)(1-\xi s)^n. \end{aligned} \tag{2.61}$$

This means that the effect of Bernoulli sampling can be taken into account by simply scaling the argument of the ideal generating function, that is,

$$q(s,\xi;T) = q(\xi s,1;T). \tag{2.62}$$

Since the mean of the distribution is $-(dq/ds)$ evaluated at $s = 0$, as might be expected this result implies that the mean will be reduced by the efficiency factor.

2.6 Summary

- In this chapter we have introduced the notion of distributions that measure the fraction of populations in an ensemble of similar ones that contain a specific number of individuals.
- We have given examples of such distributions and defined generating functions that can simplify calculation of their moments and other properties.
- We have generalised the initial concepts to include more than one variable and used this approach to introduce the notion of correlation and discrete processes that evolve with time, such as the number of individuals in an evolving population.
- Series of events are a different kind of discrete process and we have indicated that they can be generated by individuals leaving a population that is evolving in time.
- We have described two different ways that such a process can be characterised, namely, the number of events in fixed time intervals and the time between events.

Problems

2.1 The probability distribution governing the number of individuals in a population is negative binomial of index unity. What is the probability of finding fewer than M individuals present at any given time?

2.2 Calculate the variance of a population characterised by the Laguerre distribution (2.11) when the index $\alpha = 1$. Find the mean and variance of the distribution for arbitrary α when $N_1 = 0$ and also when $N_2 = 0$.

2.3 A population contains two types of individuals. The number of each type is governed by an independent thermal distribution. What is the form of the probability distribution of the total number of individuals (a) when the mean values of the two types of individuals present are the same and (b) when they are different?

2.4 The probability of finding no individuals present in a stable population, governed by the generating function (2.43) with index $\nu < 1$, is the same as that for a second population that is known to be Poisson distributed. Show that (a) the probability of finding a single individual present in the stable population is less than that of finding one present in the Poisson population and (b) the probability of finding two individuals present in the stable population is greater than that for the Poisson population if the mean of the Poisson population is less than $\nu/(1 + \nu)$.

2.5 The number of events in an integration time T is found to be approximately governed by a thermal distribution for sufficiently short time intervals. Show that at large event rates the probability distribution of the time between events is approximately an inverse power law.

Further Reading

M.S. Bartlett, *An Introduction to Stochastic Processes,* Cambridge University Press, 1966.

D.R. Cox and P.A.W. Lewis, *The Statistical Analysis of Series of Events,* Methuen, 1966.

D.R. Cox and H.D. Miller, *The Theory of Stochastic Processes,* Chapman and Hall, 1995.

W.B. Davenport and W.L. Root, *An Introduction to the Theory of Random Signals and Noise,* IEEE Press, 1987.

3

Markovian Population Processes

3.1 Introduction

The simplest random processes are those in which the value of the variable at a given time is independent of its value at any other time. The joint probabilities of finding values of the variable at different times are then a product over the first-order or *single-fold* probability densities:

$$P(x_1;t_1, x_2;t_2, \ldots x_n;t_n) = P(x_1;t_1) \times P(x_2;t_2) \times \ldots P(x_n;t_n). \tag{3.1}$$

Thus such processes are completely described by these single-fold probability densities. It is not difficult to find examples of this type of process when time is discrete; for example, the toss of a coin at stated times yielding a sequence of heads and tails. However, in the case of most physical systems, the value of a variable at one time is usually related to its value at one or more previous times. One intensively studied example is that of the motion of a particle that is undergoing random changes in direction due to collisions: the position of the particle at any given time is evidently dependent on its previous trajectory. We have seen in the last chapter that in the case of random variables this kind of relationship is expressed in the form of *correlation* between functions of the variable at different times.

The simplest kind of process that can be used to model correlated behaviour is one in which the value of the random variable at a given time is dependent on its value at any one previous time. In stochastic terms, this means that the joint distribution of the variable at an ordered set of times can be expressed as a product of the first-order probability at the first time multiplied by the conditional probabilities at successively later times, that is,

$$\begin{aligned} P(x_1;t_1, x_2;t_2, \ldots x_n;t_n) = P(x_1;t_1) \times P\left(x_2;t_2 \middle| x_1;t_1\right) \\ \times P\left(x_3;t_3 \middle| x_2;t_2\right) \ldots P\left(x_n;t_n \middle| x_{n-1};t_{n-1}\right). \end{aligned} \tag{3.2}$$

Here $t_r > t_{r-1}$ and $P(x_r;t_r|x_{r-1};t_{r-1})$ are interpreted as the *conditional* probability of finding that the variable x has the value x_r at time t_r given that it had the

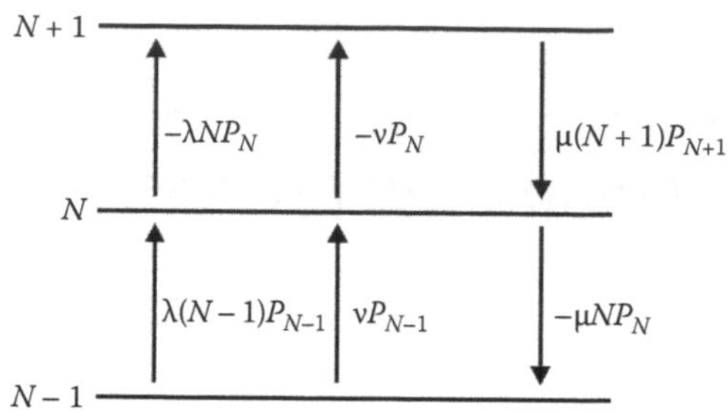

FIGURE 3.1
Transitions between different population levels in the birth–death–immigration process.

value x_{r-1} at the earlier time t_{r-1}. Random variables with the property (3.2) are called *Markov* processes after the Russian mathematician who first studied them. Strictly speaking, Equations (3.1) and (3.2) describe the properties of *zero* and *first*-order Markov processes, respectively, since the concepts can be extended to cases where the value of the variable at a given time is related to its value at zero, one or many previous times.

In the case of discrete variables, the simplest first-order Markov processes involve changes in an integer N by unity due to random addition or subtraction during a very small time interval δt. In the following sections we shall consider three processes of this kind that play a fundamental role in the theory of population statistics. These are *births* and *deaths*, the average numbers of which are proportional to the number of individuals present, and *immigrations*, which are independent of the existing number of individuals. Although the terminology we use here is motivated by the familiarity with population statistics in biological systems, it is worth mentioning the connection with processes described by similar mathematics in the field of quantum optics and atomic physics. The state of these systems is often described in terms of the emission and absorption of a number of energy *quanta*. Thus births are the analogue of stimulated emission and deaths are the analogue of absorption. The analogue of spontaneous emission is immigration, but quantum mechanics requires that this must have a rate constant equal to the stimulated (birth) rate in the same population of photons or atoms. Changes of this type are often represented as transitions between adjacent energy levels in diagrams of the type shown in Figure 3.1 and this is also a convenient representation for population processes.

3.2 Births and Deaths

Consider the effect of a simple birth process on a population of N individuals. For the population to be of size N at time $t + \delta t$, either it was of size N at time t and no birth occurred in the short space of time δt, or it had the value

$N-1$ at time t and a birth did occur in the succeeding interval. By choosing δt to be *sufficiently* small we can ensure that the probability of more than one birth occurring is negligible. Now the probability of a birth occurring will be proportional to the number present in the population. If we assume that the constant of proportionality or birth rate is λ then the probability of N increasing to $N+1$ in the interval δt will be $\lambda N\delta t$. It follows that the probability of no birth occurring is $1-\lambda N\delta t$. Similarly, the probability of $N-1$ increasing to N in the interval δt is $\lambda(N-1)\delta t$. Suppose that the probability of finding N at time t is $P_N(t)$ then

$$\begin{aligned} P_N(t+\delta t) &= P_N(t)\times \Pr\{\text{no birth in } (t,t+\delta t)\} + P_{N-1}(t)\times \Pr\{\text{one birth in } (t,t+\delta t)\} \\ &= P_N(t)\times(1-\lambda N\delta t) + P_{N-1}(t)\times\lambda(N-1)\delta t. \end{aligned} \tag{3.3}$$

This may be expressed in the form

$$\left[P_N(t+\delta t)-P_N(t)\right]/\delta t = -\lambda N P_N(t) + \lambda(N-1)P_{N-1}(t). \tag{3.4}$$

Taking the limit as the interval δt approaches zero and dropping t from the probability distribution for conciseness we find

$$\frac{dP_N}{dt} = -\lambda N P_N + \lambda(N-1)P_{N-1}. \tag{3.5}$$

The solution of this equation is the probability density P_N at time t. However, the solution will be subject to a given value or *boundary condition* at some other time, for example, if there were M individuals present at time $t=0$, then

$$P_N(0) = \delta_{NM}. \tag{3.6}$$

It will also have to satisfy the requirement that the total probability should be unity at all times, that is, the normalisation condition

$$\sum_{N=0}^{\infty} P_N(t) = 1. \tag{3.7}$$

It is evident from Equation (3.6) that the solution of Equation (3.5) should be interpreted as the *conditional* distribution $P(N;t|M;0)$ that N individuals are present at time t when there were exactly M present initially. Thus the probability density at a given time is determined by its behaviour at only *one* previous time and the higher-order statistics will exhibit the factorisation structure (3.2) of a first-order Markov process. The full solution of Equation (3.5) will be included in the treatment of a more general problem in the next chapter, but the evolution of the mean value of the population can be obtained rather simply by multiplying both sides of the equation by N and summing over N:

$$\frac{d\langle N(t)\rangle}{dt} = \lambda\langle N(t)\rangle. \tag{3.8}$$

Solving this with the boundary condition (3.6) gives the *mean* number in the population at time t when there were exactly M individuals present initially:

$$\langle N(t|M;0)\rangle = M\exp(\lambda t). \tag{3.9}$$

This shows that a population subject only to births grows exponentially without limit (Figure 3.2a). The exponential nature of this growth arises from the fact that the number of births is proportional to the number in the existing population.

At this point we might anticipate that the exponential increase of population predicted by Equation (3.9) could be offset by a *death* process that removes individuals from the population. If we assume that the constant of proportionality or death rate for this process is μ, then the probability of $N+1$ decreasing to N due to a death in the small interval δt will be $\mu(N+1)\,\delta t$. On the other hand, the probability of *no* death occurring is $1 - \mu\delta tN$. Equations analogous to (3.3) and (3.4) can now be written for the process and after

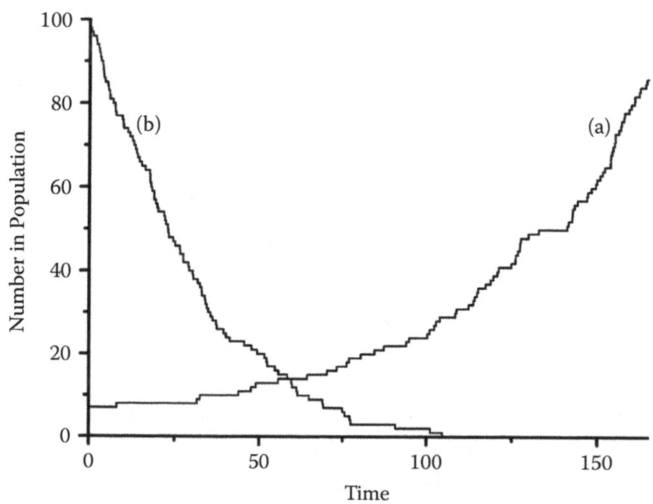

FIGURE 3.2
(a) A process with just births, $\lambda = 0.02$; (b) a process with just deaths, $\mu = 0.03$.

making δt small we obtain the following equation governing the evolution of the statistics of a population subject only to deaths

$$\frac{dP_N}{dt} = -\mu N P_N + \mu(N+1)P_{N+1}. \tag{3.10}$$

As before, we can multiply this equation by N and sum over N to obtain a simple differential equation that determines the evolution of the mean:

$$\frac{d\langle N(t)\rangle}{dt} = -\mu\langle N(t)\rangle. \tag{3.11}$$

The solution of this equation subject to the population number being M at time $t = 0$ is

$$\langle N(t|M;0)\rangle = M\exp(-\mu t). \tag{3.12}$$

Thus, as expected, the average size of the population decreases due to deaths (Figure 3.2b). In practice, of course, a single population will become *extinct* when all the individuals present initially have died since there is no mechanism for replacement included in the process (3.10). The result (3.12) for the average number of individuals present tells us that for an 'ensemble' of populations subject only to deaths, the time to extinction is typically of the order of μ^{-1}, that is, the inverse death rate, but that due to the random nature of the process, some populations will survive longer than others.

When both births and deaths are taking place Equations (3.5) and (3.10) are replaced by

$$\frac{dP_N}{dt} = -(\lambda+\mu)NP_N + \lambda(N-1)P_{N-1} + \mu(N+1)P_{N+1}. \tag{3.13}$$

Multiplying this equation by N and summing over N as before gives an equation analogous to Equations (3.8) and (3.11), and if there are M individuals present initially we find that the mean of the population at time t later is

$$\langle N(t|M;0)\rangle = M\exp[(\lambda-\mu)t]. \tag{3.14}$$

It is clear that except for the special case when the birth and death rates are equal, the mean value of the population either grows without limit or decays to zero. Thus the presence of both births and deaths does not ensure the stability of the population except in very special circumstances. In fact stability is most readily achieved by the addition of a steady stream of individuals that arrive independent of the number currently present in the population.

3.3 Immigration and the Poisson Process

By convention we use the term *immigration* to mean the process by which individuals are added to a population at a rate that is independent of the population present. Thus if we assume that the constant of proportionality or immigration rate is ν, then the probability of N increasing to $N + 1$ in the interval δt will be simply $\nu\delta t$. It follows that the probability of no immigration occurring is $1 - \nu\delta t$. Similarly, the probability of $N - 1$ increasing to N in the interval δt is $\nu\delta t$. Suppose again that the probability of finding N at time t is $P_N(t)$, then following the procedure adopted in the last section we find that

$$\frac{dP_N}{dt} = -\nu P_N + \nu P_{N-1}. \tag{3.15}$$

Multiplying by N as before and summing gives the following equation for the mean population:

$$\frac{d\langle N(t)\rangle}{dt} = \nu. \tag{3.16}$$

Solving this subject to (3.6) gives the solution

$$\langle N(t|M;0)\rangle = M + \nu t. \tag{3.17}$$

This shows that the average number in a population subject to immigration will grow *linearly* with time (Figure 3.3a). This result is reassuring because

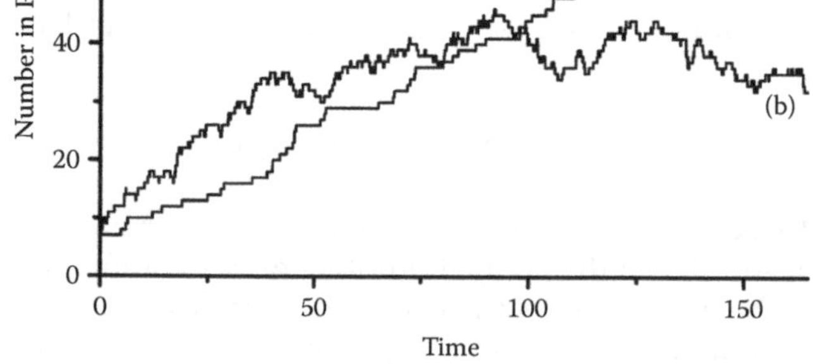

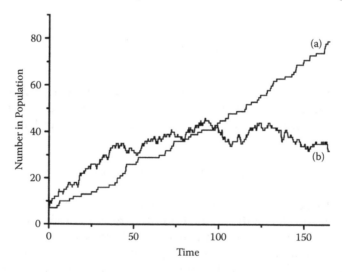

FIGURE 3.3
(a) A process with just immigrations, ν = 0.4; (b) immigrations and deaths, ν = 1, μ = 0.025.

our initial assumption was that there was a steady stream of immigrants. The question now arises as to whether, unlike the birth process, immigration could stabilise a population that would otherwise become extinct. The simplest model is one with only deaths and immigrants, and the evolution equation for the statistics of this process can be constructed from Equations (3.10) and (3.15):

$$\frac{dP_N}{dt} = -\mu N P_N + \mu(N+1)P_{N+1} - \nu P_N + \nu P_{N-1}. \tag{3.18}$$

Multiplying this equation by N and summing as before gives an equation for the mean that is now of the form:

$$\frac{d\langle N(t)\rangle}{dt} = \nu - \mu\langle N(t)\rangle. \tag{3.19}$$

It is not difficult to find the solution of this equation that satisfies Equation (3.6):

$$\langle N(t|M;0)\rangle = \bar{N} + (M - \bar{N})\exp(-\mu t), \tag{3.20}$$

where

$$\bar{N} = \frac{\nu}{\mu}. \tag{3.21}$$

It can be seen (Figure 3.3b) that in the long time limit $\mu t >> 1$ the mean of the population predicted by Equation (3.20) approaches a time-independent value, $\bar{N}$, given by Equation (3.21). This provides strong evidence that immigration stabilises the process leading eventually to a stationary equilibrium solution. Note that a complete proof of this requires the full solution of Equation (3.18) for the time evolution of the probability distribution. The method by which this can be accomplished will be described for a more general problem in the next chapter.

If, on the basis of the result (3.20), we assume that a stationary equilibrium solution does indeed exist, then eventually the solution of Equation (3.18) will become time independent and we can examine the nature of its asymptotic behaviour by setting the left-hand side of Equation (3.18) equal to zero. By inspection, the right-hand side is zero if

$$NP_N = \bar{N}P_{N-1}. \tag{3.22}$$

Here $\bar{N}$ is the equilibrium mean defined by Equation (3.21). It is not difficult to deduce from this recurrence relation that

$$P_N = \frac{\bar{N}^N}{N!}P_0. \tag{3.23}$$

Applying the normalisation condition (3.7) gives $P_0 = \exp(-\bar{N})$ so that equilibrium fluctuations in a population subject to deaths and immigration are governed by a *Poisson distribution* (see Chapter 2, Figure 2.1):

$$P_N = \frac{\bar{N}^N}{N!}\exp(-\bar{N}). \tag{3.24}$$

As we have seen in Chapter 2, the normalised factorial moments of this distribution are all equal to unity, that is,

$$N^{[r]} = \frac{\langle N(N-1)(N-2)\ldots(N-r+1)\rangle}{\bar{N}^r} = 1, \tag{3.25}$$

for all $r \geq 1$. Moreover, the ratio of the variance to the mean, or Fano factor, is also unity:

$$F = \frac{\operatorname{var} N}{\bar{N}} = 1. \tag{3.26}$$

The Poisson distribution (3.24) is usually encountered in the context of series of uncorrelated, random events. However, in this case it is the single-fold equilibrium distribution of a process that exhibits correlated fluctuations. This can be demonstrated without solving for the full time-dependent solution by noting that the bilinear moment or *number correlation function* can be expressed in terms of the conditional solution (3.20) for the mean:

$$\langle N(0)N(\tau)\rangle = \sum_{N=0}^{\infty}\sum_{M=0}^{\infty} NMP(N;\tau|M;0)P_M(0) = \sum_{M=0}^{\infty} MP_M(0)\langle N(\tau|M;0)\rangle. \tag{3.27}$$

For a stationary process, $P(N;\tau|M;0)$ is to be interpreted as the probability of finding N individuals present in the population at time τ given that there were M present initially. The sum over N is just the mean of the population at time τ given that exactly M individuals were present initially. The sum over M is the average over an ensemble of possible initial states of the equilibrium population. Thus substituting from (3.20) and summing over M using the result (3.24) for the equilibrium distribution at $t = 0$, $P_M(0)$, gives

$$\langle N(0)N(\tau)\rangle = \bar{N}^2[1-\exp(-\mu\tau)] + \langle M^2\rangle\exp(-\mu\tau) \tag{3.28}$$

From result (3.25), we find that $\langle M^2\rangle = \langle M(M-1)\rangle + \langle M\rangle = \bar{N}^2 + \bar{N}$. Therefore the normalised correlation function has the form (Figure 3.4)

$$G(\tau) = \frac{\langle N(0)N(\tau)\rangle}{\bar{N}^2} = 1 + \frac{1}{\bar{N}}\exp(-\mu\tau). \tag{3.29}$$

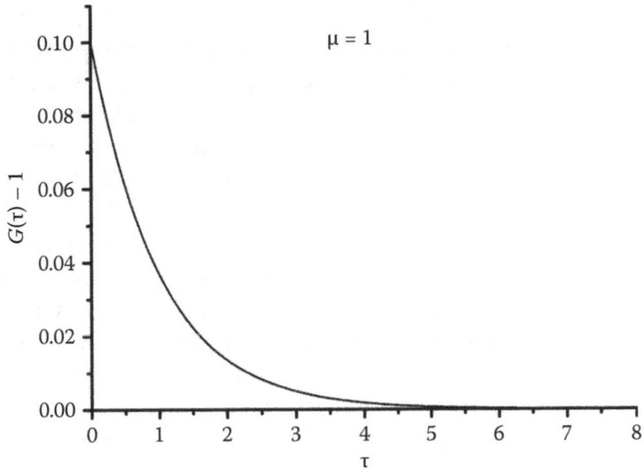

FIGURE 3.4
Correlation function for a process with deaths and immigrations. Here, the death rate is unity and the immigration rate is 10, giving an equilibrium distribution with a mean of 10 individuals in the population.

This result shows that a population subject to deaths and immigration exhibits correlated Poisson-distributed fluctuations on a timescale of the order of the inverse death rate $1/\mu$. Note, however, that the correlation coefficient or degree of correlation $G^{[2]}(\tau) - 1$ becomes small for large populations and that the fluctuations in number are uncorrelated at large separation times.

3.4 The Effect of Measurement

In the last two sections we have derived equations that describe how the statistics of populations subject to births, deaths and immigrations evolve and have calculated some simple statistical properties of the population predicated on the concept of instantaneous measurement. In practice the process of counting the population will take a finite time. If the duration of this time, though not instantaneous, is short compared with the fluctuation time of the population, we might expect real data to provide a fair test of theoretical predictions and allow parameters such as λ, μ and ν to be determined. However, it turns out that the validity of this conjecture depends upon exactly how the population is monitored. Moreover, in practice the counting time could be considerably greater than the fluctuation time so that the population will evolve during the course of the measurement and the observed statistics will be different from the instantaneous ones even if the basic model for the population process is correct.

We shall consider two kinds of monitoring processes. In the first, the population is sampled over finite time intervals and the population is unaffected by the measurement. We shall call this *internal monitoring*. In this case our mathematics just has to take account of the change in the population during a sample period. In the second monitoring process, individuals that have been counted are *removed* from the population. In addition to the effect of the finite time interval required for the count to be made, in this case the measurement process also affects the way that the population evolves. This counting model is appropriate if the count of an individual causes its death: a phenomenon that can occur in some biological measurements and is commonly encountered in quantum measurements such as photodetection. Indeed, it provides a simple classical example of a situation where measurement perturbs the system being measured—a particularly important effect if the population size is small. Another important feature of this second monitoring scheme, however, is that it can be used to describe the counting of individuals leaving a population. Thus the mathematical characterisation of the combined population evolution and measurement processes can be used to model the generation and counting of a *series of events*. We shall therefore call this second measurement scheme *external monitoring*. Note that the number of individuals counted after they have left the population may be reduced further by ineffective counting methods. Limited detector efficiency is often an important consideration in photon counting, for example. However, this may be taken into account by scaling the argument of the generating function for the ideal case as discussed at the end of Section 2.5 and we shall not consider this additional refinement here.

Since the evolution of the population and the number of counted individuals are intimately related, in order to describe the monitoring process we shall introduce the joint distribution $P(n,N;t) \equiv P_{n,N}(t)$ of there being N individuals in the population at a given time, having counted n during the preceding interval of duration t (see Figure 3.5). This quantity will be conditional on the distribution of the population at the start of the interval, for example Equation (3.6), and also on the fact that $n = 0$ at this time. Consider the change in $P(n,N;t)$ during the ensuing increment of time δt. In order to allow for imperfect counting we shall introduce a new parameter η that represents the rate at which counts are registered. The probability of a count being registered in δt is therefore $\eta N \delta t$ where, again, we shall assume that

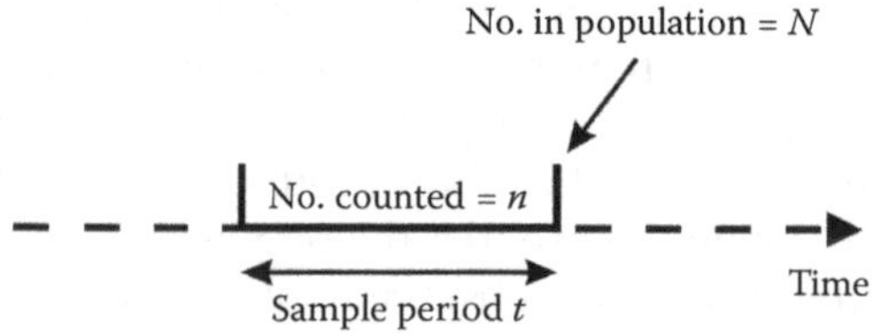

FIGURE 3.5
Monitoring process.

the time increment is so small that the probability of registering more than one count can be neglected. An equation for $P(n,N;t) = P_{n,N}(t)$ can now be derived following the pattern of earlier sections but also including the counting mechanism. In the case of internal monitoring of a population subject to deaths and immigration, governed in the absence of counting by Equation (3.18), we arrive at the result

$$\begin{aligned}\frac{dP_{n,N}(t)}{dt} &= -\mu N P_{n,N}(t) + \mu(N+1)P_{n,N+1}(t) - \nu P_{n,N}(t) + \nu P_{n,N-1}(t) \\ &\quad - \eta N P_{n,N}(t) + \eta N P_{n-1,N}(t).\end{aligned} \tag{3.30}$$

The first line of this equation is the same as (3.18) apart from the additional label n. The final term on the second line describes the transition from $n - 1$ to n counts when a count is made with N present in the population or cavity, whilst the penultimate term refers to the situation where no count is made in this situation. Similarly, for the case of external monitoring we find

$$\begin{aligned}\frac{dP_{n,N}(t)}{dt} &= -\mu N P_{n,N}(t) + \mu(N+1)P_{n,N+1}(t) - \nu P_{n,N}(t) + \nu P_{n,N-1}(t) \\ &\quad - \eta N P_{n,N}(t) + \eta(N+1) P_{n-1,N+1}(t).\end{aligned} \tag{3.31}$$

In this case the last term on the second line takes account of the decrease in the population that occurs when a count is made. As mentioned earlier, Equations (3.30) and (3.31) must be solved subject to two conditions at $t = 0$: the initial distribution of the population, for example, Equation (3.6), and $n = 0$. The solution must also reduce to P_N when summed over n and normalise to unity when summed over n and N.

By convention, the counting or sample time is usually denoted T and the distribution $p_n(T)$ of counts in this interval may be expressed in terms of the solution of Equation (3.31) as

$$p_n(T) = \sum_{M=0}^{\infty} P_M(0) \sum_{N=0}^{\infty} P_{n,N}\left(T \middle| n = 0, N = M; T = 0\right). \tag{3.32}$$

In this equation we have shown explicitly the average over the initial state of the population. The population distribution itself is obtained by setting $\eta = 0$ in Equation (3.31) and then solving the reduced Equation (3.18) or from the general solution of Equation (3.31) through summing over n:

$$P_N(t) = \sum_{n=0}^{\infty} P_{n,N}(t). \tag{3.33}$$

More general equations of the type (3.30) and (3.31) will be fully solved in the next chapter. However, it is possible to derive some useful properties of the counting distribution by following the approach of previous sections. Thus if the equations are multiplied by n and then summed over n and N, we obtain, after setting $t = T$ and using Equation (3.20), the equation

$$\frac{d\langle n\rangle}{dT} = \eta\langle N(T|M;0)\rangle = \eta\bar{N} + \eta(M-\bar{N})\exp(-\mu T). \tag{3.34}$$

This equation can now be averaged over the initial population distribution, which reduces the right-hand side to $\eta\bar{N}$ in equilibrium. Bearing in mind that there are no counts at $T = 0$, the mean number of counts simply increases linearly with sample time for both kinds of monitoring:

$$\langle n\rangle = \eta\bar{N}T. \tag{3.35}$$

Note, however, that in the case of external monitoring the population mean is given by

$$\bar{N} = \nu/(\mu+\eta) \tag{3.36}$$

rather than Equation (3.21) because of the additional 'deaths' caused by counting or by individuals leaving the population.

The correlation between n and N can be found using the same approach. In the case of internal monitoring, by multiplying Equation (3.30) with nN we obtain after summing over n and N:

$$\frac{d\langle n_i N\rangle}{dT} = -\mu\langle n_i N\rangle + \nu\langle n\rangle + \eta\langle N^2\rangle. \tag{3.37}$$

We have used the subscript n_i to denote the internally monitored counts. Substituting for $\langle N^2\rangle$ and $\langle n\rangle$ from Equations (3.25) and (3.35) gives

$$\frac{d\langle n_i N\rangle}{dT} = -\mu\langle n_i N\rangle + \nu\eta\bar{N}T + \eta\left(\bar{N}^2 + \bar{N}\right). \tag{3.38}$$

This equation must be solved subject to the condition that the correlation vanishes at $T = 0$ since the initial number of counts is zero. Thus the number-count correlation function is

$$\langle n_i N\rangle = \langle n_i\rangle\bar{N}\left\{1 + \frac{1}{\nu T}[1-\exp(-\mu T)]\right\}. \tag{3.39}$$

An equation describing the evolution of the second factorial moment of the counting distribution obtained by internal monitoring can be obtained by multiplying (3.30) by $n(n-1)$ and again summing over n and N. This gives

$$\frac{d\langle n_i(n_i-1)\rangle}{dT} = 2\eta\langle n_i N\rangle. \tag{3.40}$$

Recalling that in Equation (3.39) $\langle n_i \rangle = \eta \bar{N} T$ and that at $T = 0$ the second moment of the counting distribution must vanish, we obtain for the normalised factorial moment

$$n_i^{[2]}(T) = \frac{\langle n_i(n_i - 1)\rangle}{\langle n_i \rangle^2} = 1 + \frac{2}{\nu T}\left\{1 - \frac{1}{\mu T}[1 - \exp(-\mu T)]\right\}. \tag{3.41}$$

This function is plotted in Figure 3.6. Notice that in the limit as the counting interval goes to zero, Equation (3.41) does not reduce to the second factorial moment of a Poisson distribution (which is unity) but rather to its normalised mean square. This feature of internal monitoring will be discussed in the context of the more general model analysed in the next chapter.

As might have been expected, Equation (3.41) approaches unity when the sample time is 'sufficiently' large corresponding to the fluctuations of the population being averaged out. The Fano factor becomes a more sensitive measure in this limit, however, and reveals that the process is fluctuating despite the normalised factorial moments approaching unity. Indeed, this measure was originally introduced to detect small deviations from Poisson statistics in trains of particle detection events. If we define the counting Fano factor F_i^c as the variance of the internally monitored counting distribution divided by the mean, then from Equation (3.41) we find

$$\lim_{\mu T \to \infty} F_i^c = 1 + \frac{2\eta}{\mu}. \tag{3.42}$$

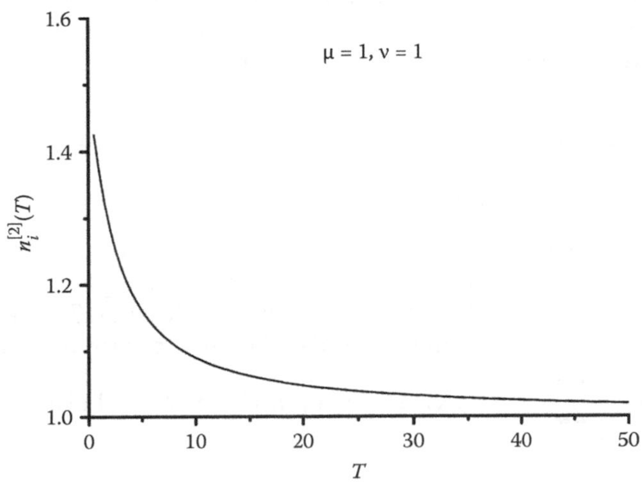

FIGURE 3.6
Second normalised factorial moment of the counting distribution obtained by internal monitoring.

This shows that the integrated fluctuations are asymptotically greater than that of a Poisson distribution with the same mean and provides a method for determining parameters of the model when the fluctuation time is much smaller than the sampling time of the monitoring process.

Now consider the case of external monitoring. Equation (3.40) is the same (with n_i replaced by n_e) but Equation (3.38) is replaced by

$$\frac{d\langle n_e N\rangle}{dT} = -(\mu+\eta)\langle n_e N\rangle + \nu\langle n_e\rangle + \eta\langle N^2\rangle - \eta\langle N\rangle. \tag{3.43}$$

As might have been expected, the death rate is augmented in this case because the counted individuals are removed from the population. However, the right-hand side is also reduced by an additional factor proportional to population. In substituting in this equation from the results of Section 3.3 the equilibrium mean of the population is now given by Equation 3.36. Following the procedure adopted earlier, the solution of Equation (3.43) can be obtained in the form

$$\langle n_e N\rangle = \eta\bar{N}^2 T = \bar{N}\langle n_e\rangle. \tag{3.44}$$

Thus, in the case of external monitoring, *the number of counts is uncorrelated with the number of individuals in the population.* Substituting from Equation (3.44) into (3.40) and integrating over T gives the result

$$n_e^{[2]}(T) = 1. \tag{3.45}$$

The second factorial moment of the counting distribution is therefore the same as that of the population distribution (3.25) irrespective of the sample time and the external counting Fano factor is unity. The more general calculation described in the next chapter confirms that, indeed, the external counting distribution for a death–immigration process is exactly Poisson.

3.5 Correlation of Counts

In this section we shall calculate the correlation between counts in different sampling intervals for a death–immigration process. In particular the correlation between counts in two *disjoint* intervals of duration T separated by an interval $\tau > T$ (Figure 3.7) can be calculated by first observing that

$$p(n,n';\tau,T) = \sum_{\substack{M,M',M'',\\ N,m}} \Pr\{n',N|M'';T\}\Pr\{m,M''|M';\tau-T\}\Pr\{n,M'|M;T\}P_M$$

$$= \sum_{M,M',M''} p(n'|M'';T)P(M''|M';\tau-T)P(n,M'|M;T)P_M. \tag{3.46}$$

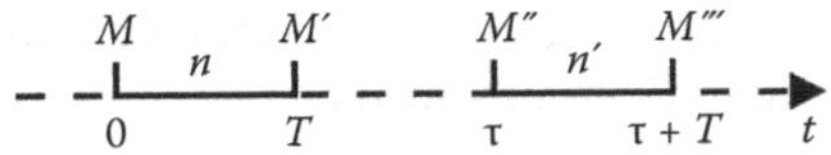

FIGURE 3.7
Correlation between different sampling intervals.

In this formula we have used a single summation sign as an abbreviation for the multiple summations from zero to infinity required. M, M', M'' represent the numbers of individuals present in the population at time $t = 0, T, \tau$, respectively as indicated in Figure 3.7. P_M is the equilibrium distribution for the population, for example (3.24), and $P(M''|M';\tau-T)$ is the probability of finding M'' in the population conditional on there being M' at a time $\tau - T$ earlier. $p(n'|M'';T)$ is the probability of having counted n' in the second interval of duration T conditional on there being M'' present in the population at time τ, whilst $P(n,M'|M;T)$ is the joint probability of counting n in the first interval with M' present in the population at the end of the interval conditional on there having been M at the beginning of the interval.

The normalised bilinear moment or correlation function of the counts is defined to be

$$g(\tau;T) = \frac{\langle n(0)n(\tau)\rangle}{\langle n\rangle^2} = \frac{1}{\langle n\rangle^2}\sum_{n=0}^{\infty}\sum_{n'=0}^{\infty} nn'p(n,n';\tau,T). \tag{3.47}$$

Multiplying $p(n'|M'';T)$ by n' and summing over this variable gives the counting mean value at time $\tau + T$ conditional on there having been M'' individuals in the population at time τ. For a stationary death–immigration process this is independent of τ and is given by Equation (3.34) replacing M with M'', that is,

$$\frac{d\langle n\rangle}{dT} = \eta\langle N(T)\rangle = \eta\bar{N} + \eta(M'' - \bar{N})\exp[-\mu T]. \tag{3.48}$$

Integrating with respect to T gives, on imposing the condition that there are no counts initially,

$$\langle n\rangle = \eta T\left[\bar{N} - \frac{M'' - \bar{N}}{\mu T}(\exp(-\mu T) - 1)\right]. \tag{3.49}$$

According to Equations 3.46 and 3.47 this must now be summed over the conditional distribution $P(M''|M';\tau-T)$, that is, the probability of finding M'' individuals in the population at time τ given that there were M' present at the earlier time $\tau - T$ (see Figure 3.7). For those terms in Equation (3.49) that are independent of M'' this just gives unity whilst for the term in M'' we use result (3.20) to obtain the reduced formula

$$g(\tau;T)=\frac{\eta T}{\langle n\rangle^2}\sum_{n,M,M'}\left\{\bar{N}-\frac{M'-\bar{N}}{\mu T}\exp[-\mu\tau](1-\exp(\mu T))\right\}np(n,M'|M,T)P_M. \tag{3.50}$$

The sums over terms not involving M' simply give a factor $\langle n\rangle$ so that using result (3.35) we obtain

$$g(\tau,T)=1-\frac{1}{\mu T}\frac{\langle nN\rangle-\langle n\rangle\bar{N}}{\langle n\rangle\bar{N}}\exp[-\mu\tau](1-\exp(\mu T)). \tag{3.51}$$

The quantity $\langle nN\rangle$ is given by Equation (3.39) for the case of internal monitoring and by (3.44) for external monitoring. Substituting these results into Equation (3.51) leads finally, for internal monitoring, to the bilinear moment

$$g_i(\tau;T)=1+\frac{\mu}{\nu}\frac{\sinh^2(\mu T/2)}{(\mu T/2)^2}\exp[-\mu\tau]. \tag{3.52}$$

This is plotted in Figure 3.8 as a function of τ for various values of T. When the sample time is much smaller than the fluctuation time (i.e., the inverse

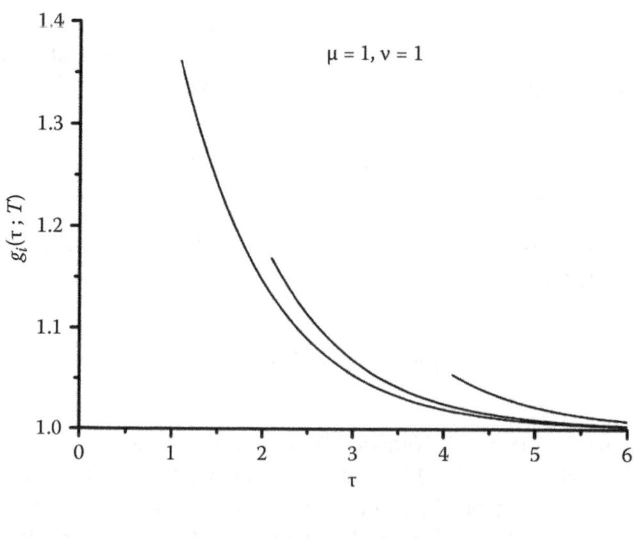

FIGURE 3.8
Correlation function for samples of duration T separated by a time interval τ, for $T = 1, 2$, and 4.

death rate) Equation (3.52) reduces to the normalised bilinear moment of the population itself, result (3.29).

For the case of external monitoring a simpler calculation gives

$$g_e(\tau;T) = 1. \tag{3.53}$$

Thus the externally measured counts are uncorrelated for all sample times. We shall show in the next chapter that in the case of a death–immigration process the number of externally measured counts is in fact Poisson distributed like the population itself. Thus the times at which individuals leave a death–immigration process constitute an *uncorrelated Poisson train of events*.

3.6 Summary

- We have shown in this chapter how the evolution of a population subject to random additions and losses of individuals can be captured by simple mathematical equations.
- We have shown that if these changes or transitions in a population occur at random times and involve only single individuals entering or leaving the population, then the number in the population at any instant of time can be related to that at any single previous instant of time, that is, the evolution of the population is a first-order Markov process.
- We have demonstrated mathematically a number of expected results; for example, that a population subject only to births grows exponentially without limit, whereas one subject only to deaths becomes extinct.
- We have found that even if a population is subject to both births and deaths, the number of individuals present will generally not stabilise but that the introduction of immigration can lead to an equilibrium state with the number in the population fluctuating about a finite mean value.
- We have examined this state for the case when only deaths and immigrations occur, and have been able to calculate its fluctuation properties and demonstrate that the probability distribution of the number in the population is Poisson.
- We have investigated two methods, internal and external monitoring, by which the statistical properties of a death–immigration population might be measured. In both cases this involves sampling over a finite time so that the population evolves during the course of the measurement, but in the case of external measurement individuals are also removed from the population when they are counted.

- We have found that, for a death–immigration process, internally monitored counts are no longer Poisson distributed and exhibit correlated fluctuations, whereas fluctuations in externally monitored counts are Poisson distributed irrespective of the sample time and are uncorrelated.

Problems

3.1 In a series of experiments on cell division a number of cells, drawn initially from a Poisson distribution, divide according to a simple birth process. After each experiment has progressed for a time t the number of individual cells is counted. Calculate the expected mean and normalised second moment of the values obtained, assuming that the number of trials is very large.

3.2 The number of photons in a cavity is attenuated by absorption according to a simple death process. Show that the moments of the initial number distribution of photons are preserved apart from a scaling of the mean.

3.3 Initially, the number of individuals in an ensemble of populations is thermally distributed. The ensuing evolution of the populations is governed by a death–immigration process. Find the mean and second moment of the distribution of individuals in the populations at time t and show how the Fano factor (variance/mean) decreases with time.

3.4 Show that, in the case of a population with only immigration and deaths, the probability distribution of the number of individuals leaving in intervals of duration T according to a simple death process remains Poisson whatever the counting interval.

3.5 The number of individuals in a population is evolving according to a death–immigration process.

a. An estimate of the mean number leaving the population is obtained by averaging the numbers counted in M consecutive time intervals of duration T. What would be the expected variance of such an estimate?

b. What would the variance of the estimated mean be in the case of internal counting?

Further Reading

M.S. Bartlett, *An Introduction to Stochastic Processes,* Cambridge University Press, 1966.

D.R. Cox and P.A.W. Lewis, *The Statistical Analysis of Series of Events,* Methuen, 1966.

D.R. Cox and H.D. Miller, *The Theory of Stochastic Processes,* Chapman & Hall, 1995.

E. Renshaw, *Modelling Biological Populations in Space and Time,* Cambridge University Press, 1991.

4

The Birth–Death–Immigration Process

4.1 Introduction

In this chapter we shall use some of the results discussed previously to write and solve a discrete process that has often been used to describe population evolution. The treatment here will serve as a template for the less familiar processes described in later chapters.

As we have seen in the last chapter, the basic birth–death process in which individuals are born and die randomly in time at a rate that is proportion to the extant population leads to a population that either grows without limit or becomes extinct according to whether the birth rate is, respectively, greater than or less than the death rate. We also showed that immigration could be used to stabilise this critical and non-stationary behaviour through study of a population subject to deaths and immigration. In this chapter we shall solve fully the more general problem that includes births as well as deaths and immigrants. Provided that the death rate is greater than the birth rate this process stabilises so that an equilibrium state is reached where the number of individuals in the population simply exhibits fluctuations about a non-vanishing finite mean number. The statistical properties of the population are then stationary (i.e., time independent) and a wide range of quantities of practical interest can be predicted. These may be used as measures to establish the validity of the model and deduce its characteristic parameters from an analysis of real data.

In addition to its role in characterising the evolution of biological populations, the birth–death–immigration process has been used to model the interaction of species in chemical reactions. It also plays a fundamental role in the theory of photon counting statistics where it was first used to describe fluctuations of the number of photons in a laser cavity operating below threshold. As mentioned in the last chapter, in this case births and deaths provide analogues of stimulated photon emission and absorption by atoms in the cavity, whilst immigration is the analogue of 'spontaneous' photon creation. The so-called thermal model of a laser below threshold is obtained when the spontaneous (immigration) and stimulated emission (birth) rates are equal as required by quantum mechanics. This special case of the birth–death–immigration process is identical to the Bose–Einstein model encountered in thermodynamics and it is commonly used to model the statistical properties of photoelectrons after detection of light that has been scattered by many independent scattering centres. Such light is found in many situations of practical interest and its properties have been exploited in the development of new optical measuring techniques and formed the basis for analyses of their performance.

More recently the birth–death–immigration process has been invoked as a model for fluctuations in the number of scattering centres encountered when electromagnetic and sound waves propagate through layers and extended regions of inhomogeneous media, or are scattered by rough surfaces. This has led to the development of *K-distributed noise* as a widely applicable, robust non-Gaussian statistical model for fluctuations in scattered scalar and vector waves that often limit the performance of optical, microwave and acoustic measuring systems operating in the natural environment.

4.2 Rate Equations for the Process

The birth–death–immigration model is a first-order Markov process that can be expressed entirely in terms of transitions in which the population changes by unity. As we have seen in Chapter 3, when individuals appear and die *randomly* in time, a stochastic representation of the problem is required and this is provided by consideration of the way in which the probability of finding a given number in the population evolves. The birth–death–immigration process is shown schematically in Figure 3.1 as transitions between the population 'levels' $N-1$, N and $N+1$. Thus births and immigration in a population of $N-1$ individuals and deaths in a population of $N+1$ individuals will all increase the probability of finding a population of N. On the other hand if any one of these processes takes place in a population of N individuals, then the probability of finding N individuals will be reduced. Following the method described in Chapter 3 we can write a rate equation for the evolution of the probability of finding N individuals in the population at time t in the form

$$\frac{dP_N}{dt} = \mu(N+1)P_{N+1} + \lambda(N-1)P_{N-1} + \nu P_{N-1} - \mu N P_N - \lambda N P_N - \nu P_N. \quad (4.1)$$

As before, μ, λ and ν are the positive proportionality constants corresponding respectively to the rates at which death, birth and immigration is taking place. This equation must be solved subject to the boundary condition at $t = 0$

and to the requirement of unit normalisation of the total probability. The general solution of Equation (4.1) is not self-evident and will be obtained later. Note, however, that a time-independent solution is possible if setting the left-hand side of the equation equal to zero gives a state that is physically realisable. A situation in which such a solution can be obtained by inspection is the *thermal model* $\nu = \lambda$, when the right-hand side is seen to be identically zero if

$$P_N = A\left(\frac{\lambda}{\mu}\right)^N. \tag{4.2}$$

This sequence of probabilities is usually called a *Bose–Einstein* distribution by the physics community, but since it forms a geometric progression it is also often referred to as a *geometric* distribution. The constant A is obtained by requiring that the total probability be normalised to unity, that is,

$$\sum_{N=0}^{\infty} P_N = \frac{A\mu}{\mu-\lambda} = 1; \quad A = 1 - \frac{\lambda}{\mu}. \tag{4.3}$$

The mean value corresponding to the distribution (4.2) can now be calculated from (4.2) and (4.3) by observing that (on setting $x = \lambda/\mu$):

$$\sum_{N=0}^{\infty} NP_N = (1-x)\sum_{N=0}^{\infty} Nx^N = (1-x)x\frac{d}{dx}\sum_{N=0}^{\infty} x^N = (1-x)x\frac{d}{dx}\frac{1}{1-x} = \frac{x}{1-x}.$$

Thus for the thermal model

$$\bar{N} = \frac{\lambda}{\mu-\lambda}. \tag{4.4}$$

This quantity must be positive and since the rate constants are also positive, the death rate in the population must be greater than the birth rate for this solution to have physical meaning.

Similarly the higher factorial moments can be calculated from Equation (4.2) by observing that

$$\begin{aligned} \langle N(N-1)(N-2)\ldots(N-R+1)\rangle &= \sum_{N=R}^{\infty} N(N-1)(N-2)\ldots(N-R+1)P_N \\ &= (1-x)\sum_{N=0}^{\infty} \frac{(N+R)!}{N!} x^{N+R}. \end{aligned} \tag{4.5}$$

The last sum on the right-hand side of this equation is $R!$ multiplied by the (inverse) binomial expansion of $1/(1-x)^{R+1}$ and so after dividing by the appropriate power of the mean we obtain the normalised factorial moments for the thermal model:

$$N^{[R]} = \frac{\langle N(N-1)(N-2)\ldots(N-R+1)\rangle}{\bar{N}^R} = R!. \tag{4.6}$$

This simple property of geometric or thermal statistics is frequently observed in optical experiments where it is found to characterise the fluctuations in photocounts obtained after the detection of laser light that has been scattered by large collections of particles or by microscopically rough surfaces.

An important aspect of the result (4.6) is that it exceeds that which would be obtained for a Poisson distributed population. This property reflects the fact that from time to time the population contains exceptionally large numbers of individuals and can be used to model *bunching* or clustering of individuals in space or time. As indicated in Chapter 2, a useful parameter that measures this property is the *Fano factor*:

$$F = \frac{\mathrm{var}\, N}{\bar{N}} = 1 + \frac{\langle N(N-1)\rangle - \langle N\rangle^2}{\bar{N}}. \tag{4.7}$$

For Poisson number fluctuations the Fano Factor is unity. For the thermal model, according to Equation (4.6), it takes the value $1+\bar{N}$, and the variance of a geometric distribution is seen to be greater than that of a Poisson distribution with the same mean value.

4.3 Equation for the Generating Function

To establish whether the solution (4.2) can be accessed from the population distribution extant at the time when the process was first activated, it is necessary to solve the full time-dependent problem. This is facilitated by using the method of *generating functions*, described in Chapter 2. A partial differential equation for the generating function of the birth–death–immigration process can be obtained from Equation (4.1) by multiplying both sides of the equation with $(1-s)^N$ and then summing over N. We now define a generating function for the process by

$$Q(s,t) = \sum_{N=0}^{\infty} (1-s)^N P_N(t). \tag{4.8}$$

Since the total probability must be unity, this quantity must satisfy the condition

$$Q(0,t) = 1 \tag{4.9}$$

for all time. Using the relationships

$$\sum_{N=0}^{\infty} N(1-s)^N P_N = -(1-s)\frac{\partial Q}{\partial s}$$

$$\sum_{N=0}^{\infty} (N+1)(1-s)^N P_{N+1} = \sum_{N=0}^{\infty} N(1-s)^{N-1} P_N = -\frac{\partial Q}{\partial s} \tag{4.10}$$

$$\sum_{N=1}^{\infty} (N-1)(1-s)^N P_{N-1} = (1-s)^2 \sum_{N=0}^{\infty} N(1-s)^{N-1} P_N = -(1-s)^2 \frac{\partial Q}{\partial s}$$

Equation (4.1) for the probability distribution can be transformed into an equation for $Q(s,t)$:

$$\frac{\partial Q}{\partial t} = -\mu s\frac{\partial Q}{\partial s} + \lambda s(1-s)\frac{\partial Q}{\partial s} - \nu s Q. \tag{4.11}$$

This linear, first-order partial differential equation is subject to the boundary condition at $t = 0$ and the normalisation condition (4.9), and can be solved by more familiar mathematical techniques than the rate Equation (4.1). First, however, we again examine the time-independent case.

The time-independent solution of Equation (4.11), if one exists, is defined by the vanishing of the time derivative on the left-hand side. Setting the right-hand side of the equation equal to zero gives

$$(\mu - \lambda + \lambda s)\frac{dQ_\infty}{ds} + \nu Q_\infty = 0. \tag{4.12}$$

Let $L = \ln Q_\infty$. Then

$$\frac{dL}{ds} = \frac{-\nu}{\mu - \lambda + \lambda s}. \tag{4.13}$$

This can be integrated exactly to give

$$L = -\frac{\nu}{\lambda}\ln(\mu - \lambda + \lambda s) + \ln C$$

$$Q_\infty(s) = \frac{C}{(\mu - \lambda + \lambda s)^{\nu/\lambda}}. \tag{4.14}$$

Here $\ln C$ is a constant of integration that is determined by the normalisation condition (4.9). This finally leads to the time-independent solution

$$Q_\infty(s) = \left(1 + \frac{\lambda s}{\mu - \lambda}\right)^{-\nu/\lambda}. \tag{4.15}$$

The moments of the distribution are given by the derivatives of this quantity evaluated at $s = 0$, as indicated in Chapter 2. Thus the mean is obtained from the first derivative:

$$\bar{N} = -\left.\frac{dQ_\infty}{ds}\right|_{s=0} = \frac{\nu}{\mu - \lambda}. \tag{4.16}$$

This result reduces to Equation (4.4) in the thermal case $\nu = \lambda$ and to the Poisson result (Chapter 3, Equation 3.21) if $\lambda = 0$. The second factorial moment can be obtained as the second derivative of Equation (4.15) at $s = 0$ and may conveniently be expressed in the normalised form

$$N^{[2]} = \frac{\langle N(N-1)\rangle}{\bar{N}^2} = 1 + \frac{\lambda}{\nu}. \tag{4.17}$$

Again, when $\nu = \lambda$ this reduces to the value of 2 predicted by Equation (4.6) for the thermal case. Calculating the higher order moments and probability densities by successive differentiation is tedious but general expressions for these quantities can be obtained more easily by observing that Equation (4.15) can be expanded about $s = 0$ and $s = 1$ as follows:

$$\left(1 + \frac{\lambda s}{\mu - \lambda}\right)^{-\nu/\lambda} = \sum_{n=0}^{\infty} \binom{n + \nu/\lambda - 1}{n} \left(\frac{-\lambda}{\mu - \lambda}\right)^n s^n, \tag{4.18}$$

$$\left(1 + \frac{\lambda s}{\mu - \lambda}\right)^{-\nu/\lambda} = \left(\frac{\mu - \lambda}{\mu}\right)^{\nu/\lambda} \sum_{n=0}^{\infty} \binom{n + \nu/\lambda - 1}{n} \left(\frac{\lambda}{\mu}\right)^n (1 - s)^n. \tag{4.19}$$

In these expressions the negative binomial coefficients are defined in terms of gamma and factorial functions in the usual way:

$$\binom{n + r}{n} = \frac{\Gamma(n + r + 1)}{n!\,\Gamma(r + 1)}. \tag{4.20}$$

The normalised factorial moments and probability distributions can now be written immediately using the formulae from Chapter 2, Equations (2.21) and (2.22):

$$N^{[R]} = \frac{\langle N(N-1)(N-2)\ldots(N-R+1)\rangle}{\bar{N}^R} = \frac{\Gamma(R + \nu/\lambda)}{(\nu/\lambda)^R\,\Gamma(\nu/\lambda)}, \tag{4.21}$$

$$P_N = \left(\frac{\lambda}{\mu}\right)^N \binom{N + \nu/\lambda - 1}{N} \left(\frac{\mu - \lambda}{\mu}\right)^{\nu/\lambda} = \binom{N + \nu/\lambda - 1}{N} \frac{\left(\lambda\bar{N}/\nu\right)^N}{\left(1 + \lambda\bar{N}/\nu\right)^{N + \nu/\lambda}}. \tag{4.22}$$

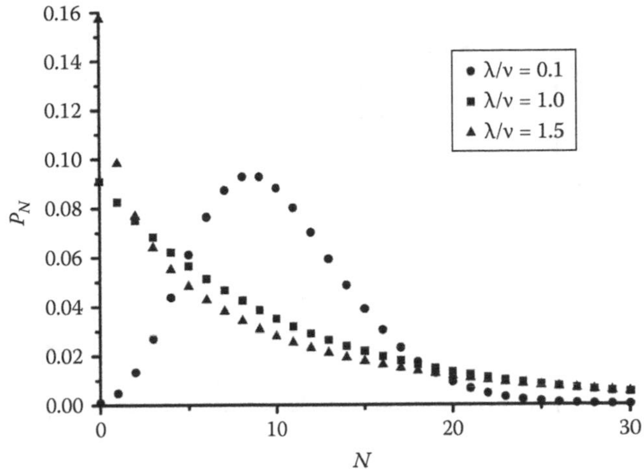

FIGURE 4.1
Equilibrium distribution for the birth–death–immigration process with a mean value of 10.

The predicted time-independent solution for the fluctuations of a birth–death–immigration process is therefore governed by the negative binomial coefficients. The distributions and normalised factorial moments given by Equation (4.22) and Equation (4.21) are plotted in Figure 4.1 and Figure 4.2 as a function of λ/ν. As the birth rate approaches zero the distribution becomes Poisson, with a peak near the mean value of the population (Figure 4.1), whereas when $\lambda = \nu$ the thermal distribution shows a monotonic decrease with increasing N. The normalised factorial moments increase monotonically with increasing birth rate from a value of unity when there are only deaths and immigration (Figure 4.2).

The aforementioned discussion is predicated on the accessibility of the time-independent solutions from given starting conditions for the process. A simple way to test this is by calculating the time evolution of the mean value of the population assuming that exactly M individuals are present at time $t = 0$. This may be accomplished by evaluating the derivative with respect to s of Equation (4.11) at $s = 0$, which leads to the first-order ordinary differential equation

$$\frac{d\langle N(t)\rangle}{dt} = (\mu - \lambda)\langle N(t)\rangle - \nu. \tag{4.23}$$

Defining the stationary mean by Equation (4.16), the solution of this equation assuming that there are M individuals present at time zero can be written:

$$\langle N(t)\rangle = \bar{N} + (M - \bar{N})\exp[-(\mu - \lambda)t]. \tag{4.24}$$

We see that the inequality $\mu > \lambda$ required for a positive mean value also ensures that at long times all memory of the initial condition is lost and

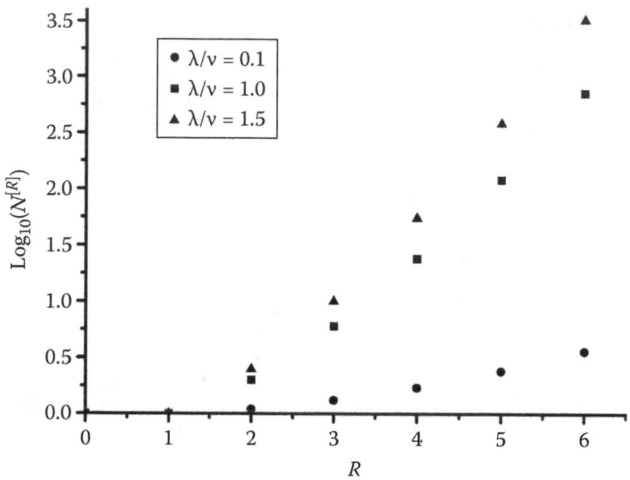

FIGURE 4.2
Normalised factorial moments for the birth–death–immigration process.

the mean approaches the time-independent result (4.16) derived earlier. Moreover, this argument applies to any initial distribution of individuals that has a finite mean value since this more general initial condition is applied by summing Equation (4.24) over the probability of finding M. This strongly suggests that Equation (4.15) is indeed a valid equilibrium solution for the birth–death–immigration problem. However, a complete proof requires calculation of the time-dependent generating function.

4.4 General Time-Dependent Solution

The time evolution of the second factorial moment can be calculated by using an approach similar to that used in calculating the evolution of the mean: differentiating Equation 4.11 twice with respect to s and then setting $s = 0$. This gives a first-order ordinary differential equation driven by the time-dependent mean value (4.24). However, an exact time-dependent solution of Equation 4.11 governing the behaviour of all moments, distributions and correlation functions can be obtained directly by the use of the method of Laplace transforms. This method is particularly suited to first-order Markov processes and so we shall describe it in this section in some detail. First define the Laplace transform of the generating function by

$$\tilde{Q}(s,p) = \int_0^\infty dt\, Q(s,t)\exp(-pt). \tag{4.25}$$

Multiplying both sides of Equation (4.11) by exp(−*pt*) and integrating over time from zero to infinity leads to the following ordinary differential equation for $\tilde{Q}$:

$$-Q(s,0)+p\tilde{Q}(s,p)=-\mu s\frac{d\tilde{Q}}{ds}+\lambda s(1-s)\frac{d\tilde{Q}}{ds}-\nu s\tilde{Q}. \tag{4.26}$$

The terms on the left-hand side of this equation are obtained through integration by parts. Equation (4.26) may be integrated over *s* using the method of integrating factors. Multiplying the equation by *R* enables it to be expressed in the form

$$R\frac{d\tilde{Q}}{ds}+\frac{p+\nu s}{\mu s-\lambda s(1-s)}R\tilde{Q}=\frac{d(R\tilde{Q})}{ds}=\frac{RQ(s,0)}{\mu s-\lambda s(1-s)}, \tag{4.27}$$

where

$$\frac{dR}{ds}=\frac{p+\nu s}{\mu s-\lambda s(1-s)}R. \tag{4.28}$$

Equation (4.28) can be integrated exactly to give the integrating factor

$$R=s^{\frac{p}{\mu-\lambda}}(\mu-\lambda+\lambda s)^{\frac{\nu}{\lambda}-\frac{p}{\mu-\lambda}}. \tag{4.29}$$

From Equation (4.27) we see that the Laplace transform of the generating function can be expressed as the integral

$$\tilde{Q}(s,p)=\frac{1}{R(s,p)}\int_0^s\frac{ds'}{s'}\frac{R(s',p)Q(s',0)}{\mu-\lambda+\lambda s'}. \tag{4.30}$$

This solution satisfies the required normalisation of the probability density embodied in Equation (4.9). This implies that the Laplace transform (4.25) must satisfy the condition

$$\tilde{Q}(0,p)=p^{-1}. \tag{4.31}$$

The required generating function is therefore the inverse Laplace transformation:

$$Q(s,t)=\int_0^s\frac{ds'}{s'}\int_{C-i\infty}^{C+i\infty}dp\exp(pt)\frac{Q(s',0)R(s',p)/R(s,p)}{(\mu-\lambda+\lambda s')}. \tag{4.32}$$

Using the explicit expression for the multiplying factor R given in Equation (4.29), the p-dependent part of the kernel of the inverse Laplace transform can be arranged in the form

$$\exp\left\{p\left[t+\frac{1}{\mu-\lambda}\ln\left(\frac{s'(\mu-\lambda+\lambda s)}{s(\mu-\lambda+\lambda s')}\right)\right]\right\}.$$

Using the well-known result that the inverse Laplace transformation of $\exp(-px)/p$ is a theta (step) function at $t = x$ the p-integral can be performed to give a delta function on the argument of this expression. Thus Equation (4.32) reduces to

$$Q(s,t)=\int_0^s ds'\delta\left(t+\frac{1}{\mu-\lambda}\ln\left(\frac{s'(\mu-\lambda+\lambda s)}{s(\mu-\lambda+\lambda s')}\right)\right)Q(s',0)\frac{(\mu-\lambda+\lambda s')^{-1+\nu/\lambda}}{s'(\mu-\lambda+\lambda s)^{\nu/\lambda}}. \quad (4.33)$$

The integral can be evaluated exactly with the help of a simple coordinate transformation leading to the following result for the time-dependent generating function:

$$Q(s,t)=\left(1+\frac{\lambda\bar{N}s}{\nu}[1-\theta(t)]\right)^{-\nu/\lambda}Q\left(\frac{s\theta(t)}{1+(\lambda\bar{N}s/\nu)[1-\theta(t)]},0\right). \quad (4.34)$$

Here, $\bar{N}$ is the time-independent mean (4.16) and

$$\theta(t)=\exp[-(\mu-\lambda)t]. \quad (4.35)$$

Note that when $t = 0$, $\theta(t) \to 1$ and the left- and right-hand sides of Equation (4.34) are identical, thereby satisfying the boundary condition at $t = 0$, whilst Equation (4.34) approaches the time-independent solution (4.15) when $(\mu - \lambda)t >> 1$ since then $\theta(t) \to 0$ and by definition $Q(0,0) = 1$. Thus Equation (4.15) is confirmed as the equilibrium solution of the birth–death–immigration problem that is approached asymptotically in the limit $t \to \infty$ whatever the initial state of the population, provided that $\mu > \lambda$.

4.5 Fluctuation Characteristics of a Birth–Death–Immigration Population

Result (4.34) is the most general solution of the birth–death–immigration problem from which the probability of finding N individuals in the population at a given time can be found conditional on there having been M present at some earlier time. Because in the case of a first-order Markov process all the multiple joint probability distributions can be expressed entirely in terms of this quantity, Equation (4.34) determines every property of the population. Thus by differentiating Equation (4.34) once and evaluating at $s = 0$ it is not difficult to check that result (4.24) for the population mean follows when exactly M individuals are present initially, that is, if

$$Q(s,0) = (1-s)^M. \tag{4.36}$$

The evolution of the second factorial moment can similarly be calculated by evaluating the second derivative of Equation (4.34) at $s = 0$. This gives

$$\langle N(t)(N(t)-1)\rangle = \langle N\rangle^2\left(1+\frac{\lambda}{\nu}\right)(1-\theta)^2 + 2\left(1+\frac{\lambda}{\nu}\right)M\langle N\rangle\theta(1-\theta) + M(M-1)\theta^2. \tag{4.37}$$

Here $\theta = \theta(t)$ is defined by Equation (4.35). It is not difficult to check that Equation (4.37) reduces to the correct result of $M(M - 1)$ in the limit of small t, and approaches the stationary second factorial moment given by Equation (4.17) when t is large.

A particularly useful property of the birth–death–immigration model is the parameter-dependent degree of fluctuation exhibited by the population. This is manifest in Figure 4.2 and in its simplest form in formula (4.17) for the normalised second moment of the distribution. A useful comparison can be made with the equivalent property of a Poisson distribution having the same mean value:

$$N^{[2]} = 1 + \frac{\lambda}{\nu} \quad \text{negative binomial distribution,} \tag{4.38}$$

$$N^{[2]} = 1 \quad \text{Poisson distribution.} \tag{4.39}$$

Another way that this comparison can be made is via the Fano factor (4.7)

$$F = 1 + \frac{\lambda}{\nu}\bar{N}. \tag{4.40}$$

Evidently fluctuations in a birth–death–immigration population exceed those of a purely random (Poisson) number of individuals with the same mean value by an amount that is determined by the ratio of the birth rate to the immigration rate. As we have already noted, this is due to the fact that from time to time the population contains exceptionally large numbers of individuals, a property that is also manifest in the *correlation* of the fluctuations. We have seen in earlier chapters that the simplest measure of correlation for the case of a stationary process is the normalised bilinear moment or number correlation function

$$G(\tau) = \frac{\langle N(t)N(t+\tau)\rangle}{\langle N\rangle^2}. \tag{4.41}$$

This can be calculated as in Chapter 3 by noting that

$$\langle N(0)N(t)\rangle = \sum_{N=0}^{\infty}\sum_{M=0}^{\infty} NMP(N;t|M;0)P_M(0) = \sum_{M=0}^{\infty} MP_M(0)\langle N(t|M;0)\rangle. \tag{4.42}$$

For a stationary process, $P(N;t|M;0)$ is to be interpreted as the probability of finding N individuals present in the population at time t given that there were M present initially. The sum over N is just the mean of the population at time t given that exactly M individuals were present initially. This can be calculated directly from Equation (4.34) using the boundary condition (4.36) and is given by result (4.24). The bilinear moment is therefore obtained by multiplying this formula by M and averaging over the stationary distribution (4.22):

$$\begin{aligned}\langle MN(t+\tau)\rangle &= \left\langle M\bar{N} + M(M-\bar{N})\exp[-(\mu-\lambda)\tau]\right\rangle \\ &= \bar{N}^2 + (\langle N^2\rangle - \bar{N}^2)\exp[-(\mu-\lambda)\tau].\end{aligned} \tag{4.43}$$

Using result (4.17) for the second moment leads, after normalisation with the square of the mean, to the result

$$G(\tau) = 1 + \left(\frac{\lambda}{\nu} + \frac{1}{\bar{N}}\right)\exp[-(\mu-\lambda)\tau]. \tag{4.44}$$

This is plotted in Figure 4.3. The quantity

$$\tau_c = (\mu-\lambda)^{-1} \tag{4.45}$$

is often called the *correlation* or *fluctuation time* of the population and expresses the average relationship between the number in the population at time $t = \tau$ as a function of the number at the earlier time $t = 0$. Formula (4.44) agrees with result (4.38) when τ is small so that $G(\tau) \to \langle N^2\rangle/\bar{N}^2$ and to unity when $(\mu - \lambda)\tau \gg 1$ showing that the number in the population is uncorrelated at times separated by much more than the characteristic fluctuation time.

A more general result for the joint probability of finding M individuals in the population initially and N at a later time t can be found by defining the generating function

$$Q(s,s';t) = \langle (1-s)^M(1-s')^N\rangle = \sum_{M=0}^{\infty}(1-s)^M P_M Q(s',t|M,0). \tag{4.46}$$

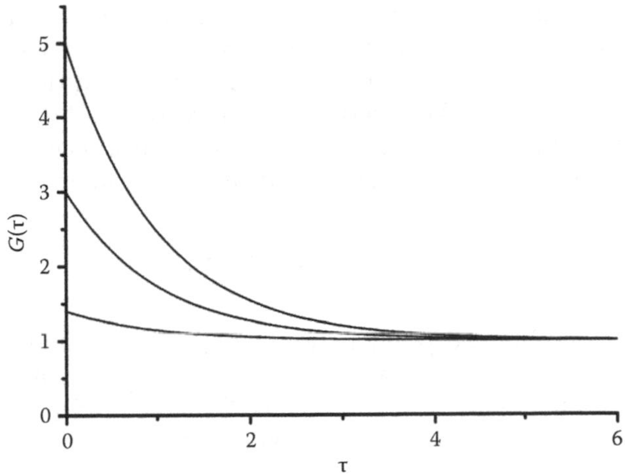

FIGURE 4.3
Correlation function of the birth–death–immigration process for $\mu = 2$, $\lambda = 1$, and three different values of ν: 0.5, 1.0 and 5.0.

This can be evaluated using result (4.34) for the conditional generating function:

$$Q(s,s';t) = \left[\left(1+\frac{\lambda\bar{N}}{\nu}s\right)\left(1+\frac{\lambda\bar{N}}{\nu}s'\right)-\frac{\lambda\bar{N}}{\nu}\left(1+\frac{\lambda\bar{N}}{\nu}\right)ss'\theta(t)\right]^{-\frac{\nu}{\lambda}}. \tag{4.47}$$

All of the joint statistical properties of the birth–death–immigration model can be calculated from this result, including the correlation function (4.44).

In the limit $\lambda \to 0$ the birth–death–immigration process degenerates into the simple death–immigration model analysed in Chapter 3. No difficulties are experienced in taking this limit in the previous formulae. For example, if we take the logarithm of result (4.14) and expand to first order in λ/μ we find

$$\ln Q_\infty = -\frac{\nu}{\lambda}\ln\left(1+\frac{\lambda s}{\mu-\lambda}\right) \approx -\frac{\nu s}{\mu}, \tag{4.48}$$

so that

$$Q_\infty = \exp(-s\nu/\mu). \tag{4.49}$$

This is the generating function for a Poisson distribution of mean $\bar{N} = \nu/\mu$:

$$P_N = \frac{\bar{N}^N}{N!}\exp(-\bar{N}). \tag{4.50}$$

As we have remarked previously, the normalised higher factorial moments of this distribution are all equal to unity. However, as we have seen in the last chapter, although the number of individuals in a death–immigration population at any time is Poisson distributed, correlated fluctuations occur due to the interaction of the immigration and death processes. According to Equation (4.44) these are characterised by a bilinear moment of the form:

$$G(\tau) = 1 + \frac{1}{\bar{N}}\exp[-\mu\tau]. \tag{4.51}$$

This is the result obtained previously in Chapter 3, Equation (3.29). Note, however, that unlike the general case (4.44) where λ is non-zero, the correlation coefficient, $G(\tau) - 1$ predicted by Equation (4.51) for the death–immigration process, vanishes if the total number of individuals in the population becomes large.

Figure 4.4 compares computer simulations of the evolution of a birth–death–immigration process and a death–immigration process with the same mean and fluctuation time. These highlight the difference in degree of fluctuation of the population numbers quantified by the formulae that we have derived earlier.

4.6 Sampling and Measurement Processes

In this section we examine the two measurement schemes introduced in the last chapter for the case when the population evolution is governed by the full birth–death–immigration process. As we have seen in Chapter 3, if we adopt schemes based on the integration intervals shown in Figure 3.5, the partial differential equation governing the counting statistics can be obtained from the one describing the instantaneous process by the addition of some extra terms. For the case of internal monitoring Equation 4.1 gives

$$\begin{aligned}\frac{dP_{N,n}}{dT} &= \mu(N+1)P_{N+1,n} + \lambda(N-1)P_{N-1,n} + \nu P_{N-1,n} - \mu N P_{N,n} - \lambda N P_{N,n} - \nu P_{N,n} \\ &\quad + \eta N P_{N,n-1} - \eta N P_{N,n}.\end{aligned} \tag{4.52}$$

Equation (4.52) describes the statistics imposed by a process in which the individuals are unaffected by being counted and the population is unperturbed by the counting process. The alternative model described in Chapter 3 involves individuals being removed or leaving the population so

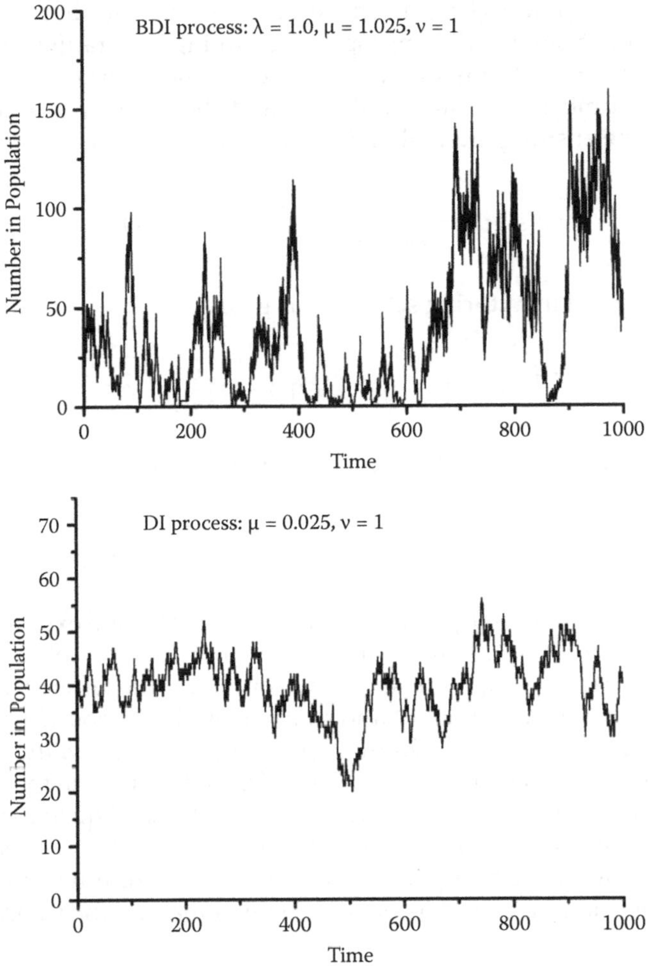

FIGURE 4.4
Comparison of fluctuations in the birth–death–immigration process (upper plot) and the death–immigration process (lower plot), both with a mean of 40.

that its evolution is changed. For this *external* monitoring scheme Equation (4.1) is modified as follows:

$$\frac{dP_{N,n}}{dT} = \mu(N+1)P_{N+1,n} + \lambda(N-1)P_{N-1,n} + \nu P_{N-1,n} - \mu N_{N,n} - \lambda N P_{N,n} - \nu P_{N,n} + \eta(N+1)P_{N+1,n-1} - \eta N P_{N,n}. \tag{4.53}$$

As we have discussed in Chapter 3, the first term in the second line of this equation describes an increase in counts following an individual being removed from the population, in contrast to the same term in Equation (4.52).

Equations (4.52) and (4.53) are most easily solved by first transforming them into equations for the appropriate counting generating functions, in this case defined in terms of two variables, s and z corresponding to the population number N at an initial time and the number of individuals n counted in the ensuing interval of duration T, respectively:

$$q(s,z;T) = \sum_{N=0}^{\infty}\sum_{n=0}^{\infty}(1-s)^N(1-z)^n P_{N,n}(T). \tag{4.54}$$

For the internal monitoring scheme this gives

$$\frac{\partial q_i}{\partial T} = -[\lambda s^2 + s(\mu - \lambda + \eta z) - \eta z]\frac{\partial q_i}{\partial s} - \nu s q_i, \tag{4.55}$$

whilst for the external monitoring scheme we obtain

$$\frac{\partial q_e}{\partial T} = -[\lambda s^2 + s(\mu - \lambda + \eta) - \eta z]\frac{\partial q_e}{\partial s} - \nu s q_e. \tag{4.56}$$

We have used the subscripts i and e to denote internal and external monitoring, respectively. The mathematical difference between the solutions of these two partial differential equations arises not only from the different coefficients of s in the first term on the right-hand side but also in the boundary condition at $T=0$, namely, $q_i(s,z;0) = (1+\lambda s/(\mu-\lambda))^{-\nu/\lambda}$, $q_e(s,z;0) = (1+\lambda s/(\mu+\eta-\lambda))^{-\nu/\lambda}$. In the case of external monitoring the loss of individuals from the population must be taken into account as an additional death rate. Equations (4.55) and (4.56) can be solved by following the technique described in Section 4.4, treating the new variable z as a parameter and T as the running time. A somewhat less laborious approach is possible by first making the transformation

$$\begin{aligned} s &= s_0 - r(z) \\ q &= l(s_0, T)\exp(\nu T r(z)). \end{aligned} \tag{4.57}$$

Now define r in such a way as to convert Equation (4.55) or (4.56) back to the form Equation (4.11) for the population itself. For the case of *internal* monitoring the appropriate choice is

$$r(z) = \frac{1}{2\lambda}\left[\mu - \lambda + \eta z - \sqrt{(\mu-\lambda+\eta z)^2 + 4\eta\lambda z}\right]. \tag{4.58}$$

This transforms Equation (4.55) into the form

$$\frac{\partial l}{\partial T} = -(\mu + \eta z - 2\lambda r)s_0\frac{\partial l}{\partial s_0} + \lambda s_0(1-s_0)\frac{\partial l}{\partial s_0} - \nu s_0 l. \tag{4.59}$$

Equation (4.59) is seen to be structurally identical to Equation (4.11) and its solution can be deduced from the equilibrium form of Equation (4.34) by simply substituting modified parameters. The solution satisfying the appropriate boundary conditions is

$$l = \left[1 - \frac{\lambda}{\mu - \lambda}\left(r - s_0 - \frac{\lambda r s_0 (1-\theta)}{\mu - \lambda + \eta z - 2\lambda r}\right)\right]^{-\frac{\nu}{\lambda}}$$
$$\theta = \exp[-(\mu - \lambda + \eta z - 2\lambda r)T]. \tag{4.60}$$

When $z = 0$ this result reduces to the generating function of the population distribution, Equation (4.15), whilst if $s = 0$ and $T = 0$ it is unity for all values of z reflecting the fact that there are no counts at the beginning of the sampling interval. A similar procedure can be followed for the case of external monitoring.

The main quantity of interest is often the count statistics governed by $q(0,z;T)$ that is obtained at $s = 0$ by setting $s_0 = r(z)$ in Equation (4.60) for example. Using Equation (4.57) it is readily shown that the results obtained for the two schemes are

$$q_i(0,z;T) = \frac{\exp(\alpha\gamma_i)}{\left[\cosh y_i + \left(1 + (y_i - \gamma_i)^2/2y_i\gamma_0\right)\sinh y_i\right]^{\alpha}} \tag{4.61a}$$

$$q_e(0,z;T) = \frac{\exp(\alpha\gamma_e)}{\left[\cosh y_e + \left(y_e/2\gamma_e + \gamma_e/2y_e\right)\sinh y_e\right]^{\alpha}} \tag{4.61b}$$

where

$$\alpha = \nu/\lambda; \quad \gamma_i = (\mu + \eta z - \lambda)T/2; \quad y_i^2 = \gamma_i^2 + \eta\lambda T^2 z$$
$$\gamma_0 = \gamma_i(z = 0); \quad \gamma_e = (\mu + \eta - \lambda)T/2; \quad y_e^2 = \gamma_e^2 + \eta\lambda T^2 z \tag{4.62}$$

and the subscripts i and e respectively denote internal and external monitoring as before. It can be seen by inspection that in both cases the result (4.61) is unity when $z = 0$ expressing the correct normalisation of the corresponding counting distributions.

The mean number of counts in a sample time are found by taking the first derivative of (4.61) with respect to z evaluated at $z = 0$:

$$\langle n_i \rangle = \eta T \frac{\nu}{\mu - \lambda}; \quad \langle n_e \rangle = \eta T \frac{\nu}{\mu + \eta - \lambda}. \tag{4.63}$$

Comparing these with the result for the mean of the internal population number (4.16) shows that, as expected, the mean count rate is just the population mean scaled linearly by the product of the counting efficiency and the integration time, except that the death rate in the case of external monitoring must be increased to take account of individuals leaving the population. This disturbs the internal population leading to a reduced value of $\bar{N}$. Indeed, it is clear from Equation (4.61) that the integrated counting statistics are somewhat different for the two kinds of monitoring processes that we have considered here. Thus, for the internally measured population the normalised second factorial moment of the counting distribution is calculated to be

$$n_i^{[2]} = 1 + \frac{\mu}{\nu\gamma_0}\left\{1 - \frac{1}{2\gamma_0}[1 - \exp(-2\gamma_0)]\right\}. \tag{4.64}$$

Here $\gamma_0 = \gamma_i(z=0) = (\mu - \lambda)T/2$. On the other hand, for the externally measured case we find

$$n_e^{[2]} = 1 + \frac{\lambda}{\nu\gamma_e}\left\{1 - \frac{1}{2\gamma_e}[1 - \exp(-2\gamma_e)]\right\}. \tag{4.65}$$

It is interesting to compare these results with the factorial moment of the population fluctuation distribution (4.17) in the short sample time limit. For internal monitoring we find

$$\lim_{T\to 0} n_i^{[2]} \to 1 + \frac{\mu}{\nu} = N^{[2]} + \frac{1}{\bar{N}} = \frac{\langle N^2 \rangle}{\bar{N}^2}. \tag{4.66}$$

Thus the normalised second factorial moment of the counting distribution based on internal sampling is different from that of the instantaneously sampled population *even for small integration times.* This may be contrasted with the short sample time limit of the external measurement

$$\lim_{T\to 0} n_e^{[2]} \to 1 + \frac{\lambda}{\nu} \equiv N^{[2]} = \frac{\langle N(N-1) \rangle}{\bar{N}^2}. \tag{4.67}$$

This slightly surprising result can be understood as follows. Consider first internal monitoring. During a period of time that is short compared with the population fluctuation time, the number of individuals will not change but a count may or may not be registered. This gives the number of counts an increased variance in comparison with that of the internal population. Now consider external monitoring. When a count is registered, in this case it is accompanied by the removal of an individual from the population, which therefore also acquires an additional and 'matching' variance leading to the identity (4.67).

In the long integration time limit, Equations (4.64) and (4.65) approach unity, but a more sensitive measure in this limit is the counting Fano factor as discussed in Chapter 3, Section 3.4. In the present case the predicted asymptotic behaviour is

$$\lim_{T\to\infty} F_i^c = 1 + \frac{2\eta\mu}{(\mu-\lambda)^2}$$
$$\lim_{T\to\infty} F_e^c = 1 + \frac{2\eta\lambda}{(\mu+\eta-\lambda)^2}. \tag{4.68}$$

These results again demonstrate the attraction of the Fano factor as a measure in situations where it is only possible to count events over intervals that are long compared with the intrinsic fluctuation times of the process under investigation. Plots of statistics corresponding to the two kinds of monitoring are shown in Figures 4.5 and 4.6.

Another feature of result (4.65) for external monitoring is that in the absence of births, the normalised second factorial moment is unity. This is in agreement with the result for a death–immigration process found in the last chapter. However, here we can obtain a more general result by setting the birth rate equal to zero in the full generating function (4.61):

$$\lim_{\lambda\to 0} q_e(0,z;T) = \exp(-\eta\bar{N}Tz) = \exp(-\langle n_e\rangle z). \tag{4.69}$$

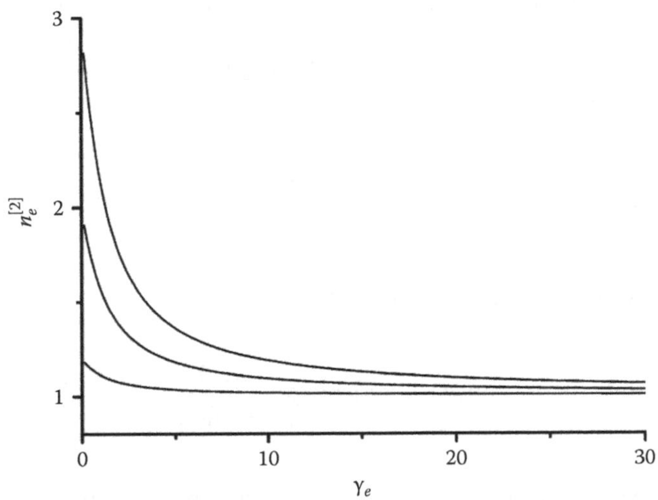

FIGURE 4.5
Normalised second factorial moment for external monitoring as a function of the parameter γ_e (which is proportional to the sample time T). The three curves are for different values of the ratio of the immigration rate to the birth rate, $\alpha = 0.5$, 1.0 and 5.0, in order of the higher curve to the lower.

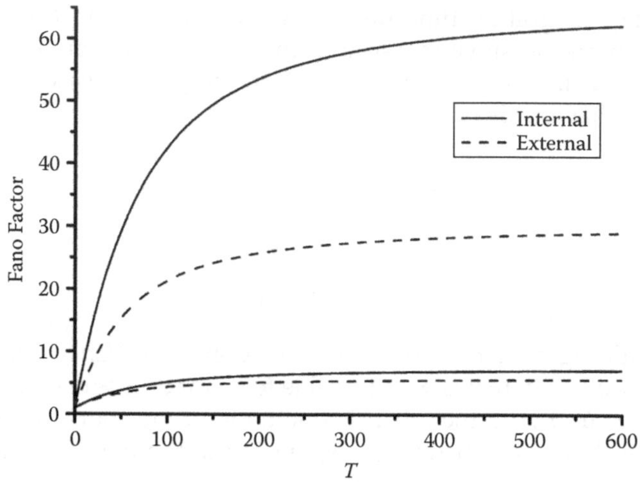

FIGURE 4.6
Comparison of the Fano factor for internal and external monitoring as a function of the sample time T. The upper curves are for a sampling rate $\eta = 0.02$, and the lower curves have $\eta = 0.002$. The other parameters are $\nu = 1$, $\mu = 1.025$ and $\lambda = 1$.

This is the generating function for a Poisson distribution, confirming the conjecture made in the last chapter that both the population and the external counting distribution are Poisson in the case of a death–immigration process.

A much more complicated result is obtained in the case of internal monitoring when the birth rate is zero:

$$\lim_{\lambda \to 0} q_i(0,z;T) = \exp\left\{\frac{-\eta\nu zT}{\mu+\eta z}\left[1+\frac{\eta z}{\mu T(\mu+\eta z)}(1-\exp[-(\mu+\eta z)T])\right]\right\}. \quad (4.70)$$

At large integration times we obtain the limit

$$\lim_{\substack{\lambda \to 0 \\ T\to\infty}} q_i(0,z;T) = \exp\left(\frac{-\eta\nu zT}{\mu+\eta z}\right). \quad (4.71)$$

Although near $z = 0$ this resembles the external monitoring result (4.69), so that the mean of both counting distributions is the same apart from the additional 'death' rate in the case of external monitoring, the higher moments are different. Indeed, it is clear from Figure 4.6 that the relative fluctuations of internally monitored counts are always greater than those obtained by external monitoring. As explained earlier, this is due to the fact that additional fluctuations are introduced by the internal monitoring process into the number of counts but not the population itself. Results for the probabilities and moments of counts corresponding to the asymptotic form (4.71) can be obtained by noting that the right-hand side is the generating function for a difference of Laguerre polynomials.

4.7 Correlation of Counts

The correlation between counts in two disjoint intervals of duration T separated by an interval $\tau > T$ (Figure 3.7) can also be calculated by following the method adopted in Chapter 3, first observing that

$$p(n,n';\tau,T) = \sum_{\substack{M,M',M'',\\ N,m}} \Pr\{n',N|M'';T\}\Pr\{m,M''|M';\tau-T\}\Pr\{n,M'|M;T\}P_M$$

$$= \sum_{M,M',M''} p(n'|M'';T)P(M''|M';\tau-T)p(n,M'|M;T)P_M. \tag{4.72}$$

The symbols in this formula have the same interpretation as described in Chapter 3, Section 3.5, and the bilinear moment or correlation function of the number fluctuations is defined as before by

$$g(\tau;T) = \frac{\langle n(0)n(\tau)\rangle}{\langle n\rangle^2} = \frac{1}{\langle n\rangle^2}\sum_{n=0}^{\infty}\sum_{n'=0}^{\infty} nn'p(n,n';\tau,T). \tag{4.73}$$

Multiplying $p(n'|M'';T)$ by n' and summing over this variable gives the counting mean value at time $\tau + T$ conditional on there having been M'' individuals in the population at time τ. For a stationary process this is independent of τ and can be obtained from Equation (4.55) or Equation (4.56) by differentiating once with respect to z and then setting s and z both equal to zero, recalling that the mean number in the population at the end of the interval is given by Equation (4.24) with $t = T$. For the internal counting process we obtain

$$\frac{d\langle n\rangle}{dT} = \eta\langle N(T)\rangle = \eta\langle N\rangle + \eta(M'' - \langle N\rangle)\exp[-(\mu-\lambda)T]. \tag{4.74}$$

Integrating with respect to T gives

$$\langle n\rangle = \eta T\left[\langle N\rangle - \frac{M''-\langle N\rangle}{2\gamma_0}(\exp(-2\gamma_0)-1)\right]. \tag{4.75}$$

According to Equations (4.72) and (4.73) this must now be summed over the conditional distribution $P(M''|M';\tau-T)$, that is, the probability of finding M'' individuals in the population at time τ given that there were M' present at the earlier time τ - T. For those terms in Equation (4.75) that are independent of M'' this just gives unity whilst for the term in M'' we again use result (4.24) to obtain the reduced formula

$$g(\tau;T)=\frac{\eta T}{\langle n\rangle^2}\sum_{n,M,M'}\left\{\langle N\rangle-\frac{M'-\langle N\rangle}{2\gamma_0}\exp[-(\mu-\lambda)\tau](1-\exp(2\gamma_0))\right\}$$
$$\times np(n,M'|M,T)P_M. \tag{4.76}$$

The sums over terms not involving M' simply give a factor $\langle n\rangle$ so that

$$g(\tau,T)=1-\frac{\langle nN\rangle-\langle n\rangle\bar{N}}{2\gamma_0\langle n\rangle\bar{N}}\exp[-(\mu-\lambda)\tau](1-\exp(2\gamma_0)). \tag{4.77}$$

When there are no births this result reduces to formula (3.51) (see Chapter 3). The quantity $\langle nN\rangle=\partial^2 q_i/\partial z\partial s\,|_{s=z=0}$ can be obtained for internal monitoring from Equation (4.55) by differentiating with respect to z and then with respect to s. After setting $s = z = 0$ we obtain the equation

$$\frac{d\langle nN\rangle}{dT}=-(\mu-\lambda)\langle nN\rangle+\nu\langle n\rangle+\eta\bar{N}+\eta\langle N(N-1)\rangle. \tag{4.78}$$

This can be integrated, noting that $\langle nN\rangle=0$ at the beginning of the interval and using the equilibrium results (4.16), (4.17) and (4.63) to give

$$\frac{\langle nN\rangle}{\langle n\rangle\bar{N}}=1-\frac{\mu}{2\nu\gamma_0}(\exp(-2\gamma_0)-1). \tag{4.79}$$

Substituting this result into Equation (4.77) leads finally to the bilinear moment of internally monitored counts

$$g_i(\tau;T)=1+\frac{\mu}{\nu}\frac{\sinh^2\gamma_0}{\gamma_0^2}\exp[-(\mu-\lambda)\tau]. \tag{4.80}$$

A similar calculation for the external monitoring scheme obtains the formula

$$g_e(\tau;T)=1+\frac{\lambda}{\nu}\frac{\sinh^2\gamma_e}{\gamma_e^2}\exp[-(\mu+\eta-\lambda)\tau]. \tag{4.81}$$

As might be expected, Equation (4.80) reduces to result (4.44) for the population itself when the counting interval is sufficiently small, that is, when

$\gamma_0 << 1$. Also, as expected, the death rate is increased in the correlation time appearing in Equation (4.81) due to the additional death rate corresponding to individuals leaving the population. However, the term of Equation (4.44) that is inversely proportional to the population mean is missing from Equation (4.81). This has the interesting consequence, noted in Chapter 3 that in the absence of births, $\lambda = 0$ the externally monitored fluctuations of the population are uncorrelated. Since according to result (4.69) the distribution of external counts is Poisson in this case, it confirms the conjecture made in Chapter 3 that the times that individuals leave a death–immigration population will form an *uncorrelated random Poisson train of events.*

The origin of the differing count correlation functions can be understood using an argument similar to that used earlier for the count moments. Suppose the two contiguous counting intervals are so short that only a single event may occur in each of them. In this situation

$$\langle nn'\rangle = \sum_{nn'} nn'p(n,n';T,T) \approx p(1,1;T,T) = \sum_{N} p(1,1;T,T|N;0), \qquad (4.82)$$

where N is the number in the population initially. In the case of internal monitoring this formula can be estimated approximately as

$$\lim_{T\to 0} g_i(T;T) \approx \frac{1}{\langle n\rangle^2}\sum_{N} \eta TN \times \eta TN \times P_N = \frac{\langle N^2\rangle}{\bar{N}^2}. \qquad (4.83)$$

On the other hand the effect of external monitoring is to reduce the population size by one for every count and so

$$\lim_{T\to 0} g_e(T;T) \approx \frac{1}{\langle n\rangle^2}\sum_{N} \eta TN \times \eta T(N-1) \times P_N = \frac{\langle N(N-1)\rangle}{\bar{N}^2}. \qquad (4.84)$$

Comparison with Equation (4.17) shows that these results are in agreement with Equations (4.80) and (4.81), that is, in the limit of small time delays, the short sample time limit of the normalised bilinear moments of the internally and externally monitored counting distributions reduce to the respective normalised factorial moments (4.66) and (4.67). In this limit they are equal to the normalised mean square and the normalised second factorial moment of the population fluctuation distribution, respectively.

Figure 4.7 compares computer simulations of the series of events generated by a birth–death–immigration process and events generated by a death–immigration process with the same mean and fluctuation time. The enhanced bunching of departures from the population with births is clearly evident by comparison with the random Poisson train of events generated by departures from the population having only deaths and immigrations.

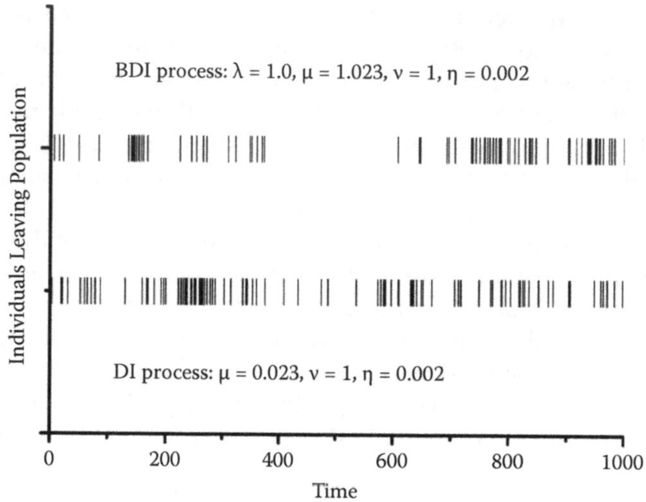

FIGURE 4.7
Comparison of individuals leaving the population for the birth–death–immigration process (upper time series) and the death–immigration process (lower time series), both with a mean of 40. These time series were derived from the populations plotted in Figure 4.4.

4.8 Summary

- In this chapter we have obtained a full solution of the birth–death–immigration process by the method of Laplace transforms.
- We have shown that when the death rate exceeds the birth rate the solution approaches an equilibrium state in which the population number fluctuates about a constant mean value determined by the rate constants.
- We have shown that the probability distribution of the equilibrium fluctuations is a negative binomial with a variance that exceeds that of a Poisson distribution with the same mean value.
- We have shown that the number correlation function decays exponentially with a time constant that is inversely proportional to the difference between the death and birth rates, and that this behaviour persists even for large populations in contrast with the simpler death–immigration process.
- We have obtained general solutions for the generating functions of the single interval internal and external counting distributions and

for the count correlation functions and have identified the origin of differences in the moments predicted for the two monitoring schemes.

- We have demonstrated the advantage of the Fano factor as a measure of deviation from Poisson statistics in situations where counting intervals are restricted to be much longer than the population fluctuation times.

Problems

4.1 The measured normalised second factorial moment of a birth–death–immigration process is found to be 2. Express the parameters of the process in terms of the mean number of individuals in the population $\bar{N}$ and its fluctuation time τ_c.

4.2 The number of individuals in an ensemble of populations evolving according to identical birth–death–immigration processes is estimated by counting the number in each population on M occasions at intervals of duration I. What would be the expected variance on an estimate of the mean?

4.3 In a series of biological experiments, the initial number of individuals in each population is Poisson distributed with mean $\bar{N}_0$. The populations then evolve according to a birth–death–immigration process.

a. Show that the population after a time t is governed by the generating function for the Laguerre distribution.

b. Calculate the distribution, factorial moments and Fano factor, and show how the parameters of the birth–death–immigration model may be recovered from measurements of the statistics of the process made at any given time of its evolution.

4.4 The mean number of photoevents registered during intervals of duration T by a photo detector outside a laser cavity operating below threshold is $\bar{n}$. The detector is not 100% efficient and counts only a fraction ξ of the events. The observed fluctuation time is τ_c. Assuming that the thermal process is a reasonable model for the evolution of the photon population within the cavity, show that a measurement of the Fano factor at long integration times can be used to determine the detector efficiency.

4.5 Obtain expressions in terms of Laguerre polynomials for the normalised factorial moments and probability distributions corresponding to result (4.71) for internal monitoring of a death–immigration process in the long integration time limit.

Further Reading

M.S. Bartlett, *An Introduction to Stochastic Processes*, Cambridge University Press, 1966.

D.R. Cox and P.A.W. Lewis, *The Statistical Analysis of Series of Events*, Methuen, 1966.

E. Jakeman and K.D. Ridley, *Modeling Fluctuations in Scattered Waves*, Taylor & Francis, 2006.

E. Renshaw, *Modelling Biological Populations in Space and Time*, Cambridge University Press, 1991.

S.K. Srinivasan, *Point Process Models of Cavity Radiation and Detection*, Charles Griffin and Co. Ltd., London, 1988.

5

Population Processes Exhibiting Sub-Poisson Fluctuations and Odd–Even Effects

5.1 Introduction

The birth–death–immigration model studied in the last chapter is conventional or *classical* in the sense that when there are many individuals present, so that the population is 'dense', a limit exists in which the population fluctuations are asymptotically characterised by the probability density of a continuous process. In the case of a Poisson distribution, the relative variance of the fluctuations vanishes as the mean becomes large and the asymptotic continuous process is governed by a delta function probability density, that is,

$$P_N \to P(N) \sim \delta(N - N_0). \tag{5.1}$$

The probability density (5.1) characterises a continuous variable that always takes the value N_0, that is, exhibits no fluctuations. Thus the high density limit of the Poisson distribution is the *narrowest* probability density possible for a continuous variable. However, it is clearly possible to envisage discrete distributions that are narrower than Poisson in the sense that their variance is less than that of a Poisson distribution with the same mean value, that is, with a Fano factor less than unity. These have no continuous counterparts but may be of particular interest in the case of small populations. On the other hand, it should be emphasised that some distributions with a Fano factor larger than unity may also have no continuous counterpart; for instance, distributions that exhibit a difference in character between even and odd numbers of individuals. Many examples of these can be envisaged in biological populations arising from twins and various kinds of pairing processes. Important even–odd effects also occur in quantum systems, for example, when pairs of particles are emitted into a population through non-linear optics effects, multiple atomic transitions or radioactive decay.

With the above comments in mind, in this chapter we investigate three simple Markov processes that lead to populations that have no continuous counterpart: two 'limited' birth–death processes that can give rise to sub-Poisson fluctuations and a death–immigrant pair model that generates even–odd effects. These will be exploited in Chapters 6–8, providing the basis for modelling and interpreting data from multi-scale and power-law systems.

5.2 A Limited Birth–Death Process: Binomial Number Fluctuations

Consider a population with the constant death rate μ and birth rate λ. Suppose that losses from the population are proportional to the extant population in the usual way but that the increase in population due to births is proportional to the difference between the population present and a fixed upper bound U. Following the method adopted in the last two chapters this *inverted* or *limited* birth–death process will be described by the rate equation

$$\frac{dP_N}{dt} = \mu(N+1)P_{N+1} - \mu N P_N - \lambda(U-N)P_N + \lambda(U-N+1)P_{N-1}. \tag{5.2}$$

Note that we must always have $U \geq N$ in order that the birth terms should be physical. In other words, the number of individuals in the population can never exceed an upper limit U, that is, $P_N = 0$ for $N > U$. This restriction also applies to the number present at the initial time. As the population number approaches the maximum value U the number of births is attenuated and vanishes if this number is reached, whereupon the death process reduces the population. However, as soon as it falls below U the possibility of births is restored and the population number can recover. It is immediately apparent that for this simple model there are no longer the problems of divergence or extinction encountered in the case of the normal birth–death process discussed in Chapter 3. In real populations, environmental factors may indeed reduce the birth rate in a population when the number of individuals becomes large. However, the model is presented here as a simple mechanism leading to sub-Poisson statistics. Some simulations are shown in Figure 5.1.

Equation (5.2) can be solved following the procedure used in the last chapter for the birth–death–immigration process by first transforming it into an equation for the generating function:

$$\frac{\partial Q}{\partial t} = -s[\mu + \lambda(1-s)]\frac{\partial Q}{\partial s} - \lambda U s Q. \tag{5.3}$$

In order to obtain this equation we have multiplied Equation (5.2) by $(1-s)^N$, summed over N using definition (4.8) for the generating function and also used relationships (4.10) (see Chapter 4). Equation (5.3) can be solved by following the Laplace transform technique described in Chapter 4, Section 4.4.

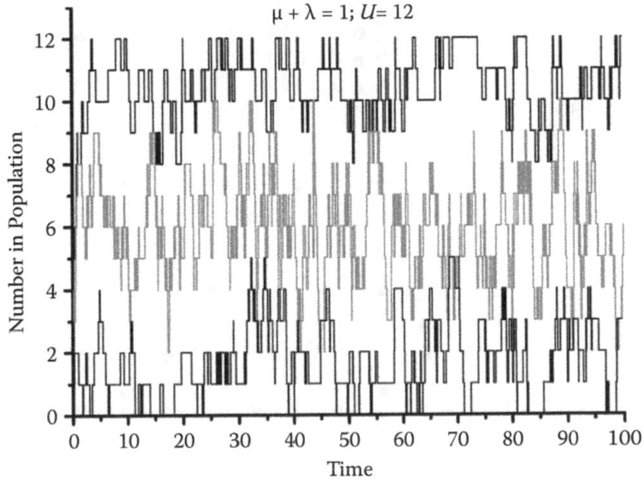

FIGURE 5.1
Population fluctuations for the limited birth–death process. The upper plot has φ = 0.9, the middle (gray) plot has φ = 0.5 and the lower plot has φ = 0.1.

However, the solution can be expedited by recognising that Equation (5.3) is obtained when we make the transformation $\lambda \to -\lambda, \nu \to \lambda U$ in Equation (4.11). Assuming that there are $M < U$ individuals present initially, the generating function for the distribution at a later time t is found to be

$$Q(s,t) = [1-(1-\theta)\quad s]^{U}\left(\frac{1-[(1-\theta)\quad +\theta]s}{1-(1-\theta)\quad s}\right)^{M}, \tag{5.4}$$

with

$$\begin{aligned}\theta(t) &= \exp\left[-(\mu+\lambda)t\right] \\ &= \lambda/(\mu+\lambda).\end{aligned} \tag{5.5}$$

The solution (5.4) satisfies the condition $Q(0,t) = 1$ for all time t that is required to maintain the normalisation of the total probability. At long times, $t >> (\mu+\lambda)^{-1}$ an equilibrium is reached with

$$Q_{\infty}(s) = (1-\quad s)^{U}. \tag{5.6}$$

Fluctuations in the population number of are then governed by the *binomial* distribution (Figure 5.2)

$$P_N = {}^{U}C_N \quad {}^{N}(1-\quad)^{U-N}. \tag{5.7}$$

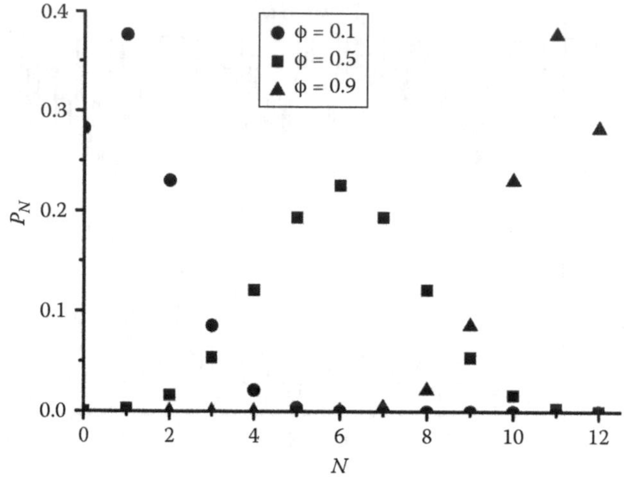

FIGURE 5.2
The binomial distribution of Equation (5.7), with $U = 12$. The three distributions correspond to the three plots in Figure 5.1.

In this last formula the binomial coefficients are defined in the usual way by ${}^{U}C_N = U!/N!(U-N)!$. The mean value of the number in the population is given by

$$\bar{N} = U \ . \tag{5.8}$$

It is not difficult to calculate the normalised factorial moments by differentiating Equation (5.6) and these are found to be less than unity (Figure 5.3):

$$N^{[r]} = \frac{U!}{U^r (U-r)!} = \left(1-\frac{1}{U}\right)\left(1-\frac{2}{U}\right)\left(1-\frac{3}{U}\right)\cdots\left(1-\frac{r-1}{U}\right). \tag{5.9}$$

The sub-Poisson nature of the population number fluctuations is confirmed by the Fano factor, which takes the value

$$F = 1 - \varphi. \tag{5.10}$$

Note that if there are no deaths, $\mu = 0, \varphi = 1$ the solution (5.4) evolves into a *number state*, that is, a population containing a *fixed* or unvarying number of individuals with zero Fano factor and generating function

$$Q_\infty(s) = (1-s)^U. \tag{5.11}$$

The correlation function of the number of fluctuations can be obtained by following the method of Chapter 3, Section 3.3, that is, by calculating the

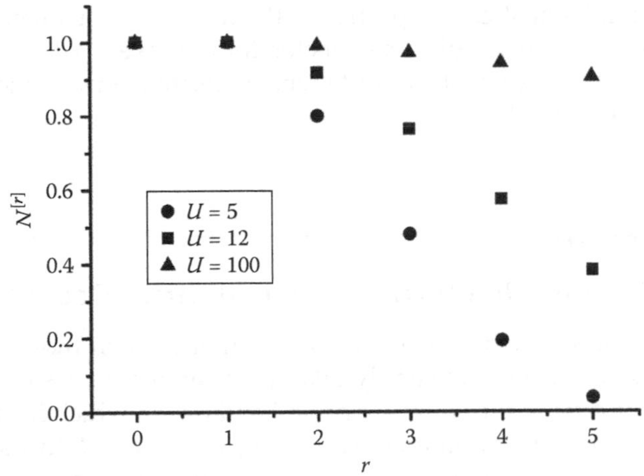

FIGURE 5.3
Normalised factorial moments for the limited birth–death process for three different values of the upper limit *U*.

conditional mean from (5.4) and using the identity (3.27)

$$\langle N(0)M(t)\rangle = \sum_{N=0}^{\infty}\sum_{M=0}^{\infty} NMP(M;t|N;0)P_N = -\sum_{N=0}^{\infty} NP_N \left.\frac{\partial Q(s;t|N;0)}{\partial s}\right|_{s=0}. \quad (5.12)$$

Here P_N is given by the equilibrium distribution (5.7) and the generating function is given by Equation (5.4). The result for the normalised correlation function is

$$G(t) = 1 + \frac{1-}{U}\theta(t). \quad (5.13)$$

Here $\theta(t)$ is defined by Equation (5.5). Result (5.13) exhibits the familiar exponential decay from the value of the normalised mean square number in the population. In the special case of a number state, $\varphi = 1$, there are no fluctuations so that $G(t) = 1$.

It is worth mentioning here that a more general result for the joint statistics of finding N individuals in the population initially and M at a later time t can be found by defining the joint generating function

$$Q(s,s';t) = \left\langle (1-s)^M (1-s')^N \right\rangle = \sum_{N=0}^{\infty} (1-s)^M P_M Q(s';t|M;0). \quad (5.14)$$

This can be evaluated using result (5.4) for the conditional generating function to give

$$Q(s,s';t) = [(1- \quad s)(1- \quad s') + \quad (1- \quad)\theta ss']^U. \quad (5.15)$$

All of the joint statistical properties of the limited birth–death model can be calculated from this result. As expected for the case $\varphi = 1$, Equation (5.15) factorises into the product of two independent number states since the population does not fluctuate.

5.3 The Effect of Monitoring a Limited Birth–Death Process

In Chapters 3 and 4 we have considered two methods for monitoring a fluctuating population: one that simply takes account of changes in the population that occur during the counting period and one in which individuals are also removed from the population when they are counted. Just as we found previously, these two monitoring schemes can be incorporated in rate equations for the joint probability of counting n individuals in the time interval T with N present in the population at the start of the interval by adding terms to the rate Equation (5.2) for the population itself. Thus for internal monitoring we obtain

$$\frac{dP_{N,n}}{dT} = \mu(N+1)P_{N+1,n} + \lambda(U-N+1)P_{N-1,n} - \mu NP_{N,n} - \lambda(U-N)P_{N,n} + \eta NP_{N,n-1} - \eta NP_{N,n}. \tag{5.16}$$

On the other hand if individuals are removed from the population when they are counted we obtain

$$\frac{dP_{N,n}}{dT} = \mu(N+1)P_{N+1,n} + \lambda(U-N+1)P_{N-1,n} - \mu NP_{PN,n} - \lambda(U-N)P_{N,n} + \eta(N+1)P_{N+1,n-1} - \eta NP_{N,n}. \tag{5.17}$$

These equations can be solved from first principals by using the generating function and the Laplace transform method described in Chapter 4. However, the solution can be expedited by recognising that Equations (5.16) and (5.17) are obtained when we make the transformation $\lambda \to -\lambda$, $\nu \to \lambda U$ in Equations (4.52) and (4.53) (see Chapter 4), respectively. Thus we find:

$$q_i(0,z;T) = \exp(-U\gamma_i)\left[\cosh y_i + \left(1 + (y_i - \gamma_i)^2/2y_i\gamma_0\right)\sinh y_i\right]^U \tag{5.18a}$$

$$q_e(0,z;T) = \exp(-U\gamma_e)\left[\cosh y_e + \frac{1}{2}\left(\frac{y_e}{\gamma_e} + \frac{\gamma_e}{y_e}\right)\sinh y_e\right]^U. \tag{5.18b}$$

Now, however, the parameters for the two cases are given by

$$\gamma_i = (\mu + \lambda + \eta z)T/2, \quad y_i^2 = \gamma_i^2 - \eta\lambda T^2 z$$
$$\gamma_e = (\mu + \lambda + \eta)T/2, \quad y_e^2 = \gamma_e^2 - \eta\lambda T^2 z. \tag{5.19}$$

The lower moments of the counting distributions can be obtained from (5.18) by evaluating the derivatives of the generating functions at $z = 0$ in the usual way or by transforming the parameters in previous results:

$$\langle n_i \rangle = \frac{\lambda\eta UT}{\mu+\lambda}, \quad \langle n_e \rangle = \frac{\lambda\eta UT}{\mu+\eta+\lambda}, \tag{5.20}$$

$$n_i^{[2]} = 1 + \frac{\mu}{\lambda U \gamma_0}\left\{1 - \frac{1}{2\gamma_0}[1 - \exp(-2\gamma_0)]\right\}, \tag{5.21}$$

$$n_e^{[2]} = 1 - \frac{1}{U\gamma_e}\left\{1 - \frac{1}{2\gamma_e}[1 - \exp(-2\gamma_e)]\right\}. \tag{5.22}$$

In these equations the parameters are defined by (5.19) with $\gamma_0 = \gamma_i(z = 0) = (\mu + \lambda)T/2$. It is clear from these results that whilst the externally monitored counts remain sub-Poisson like the population itself (but now with a Fano factor lying between a half and unity), the internally measured counts are super-Poisson with a Fano factor greater than unity (Figure 5.4). In the short sample time limit $\gamma T \ll 1$, the result (5.22) reduces to the second factorial moment of the population itself $N^{[2]}$ given by (5.9), whilst result (5.21) approaches the normalised mean square of the population distribution with the super-Poisson value of $1 + \mu/\lambda U$.

The correlation function of the internally and externally monitored counts can also be deduced from previous results. For the case of internal monitoring we find

$$g_i(\tau; T) = 1 + \frac{\mu}{\lambda U}\frac{\sinh^2 \gamma_0}{\gamma_0^2}\exp[-(\mu + \lambda)\tau]. \tag{5.23}$$

Here as before $\gamma_0 = \gamma_i(z = 0) = (\mu + \lambda)T/2$. Thus the internally monitored counts are bunched or clustered in time. For the case of external monitoring, on the other hand, the count number correlation function is calculated to be

$$g_e(\tau; T) = 1 - \frac{1}{U}\frac{\sinh^2 \gamma_e}{\gamma_e^2}\exp[-(\mu + \eta + \lambda)\tau]. \tag{5.24}$$

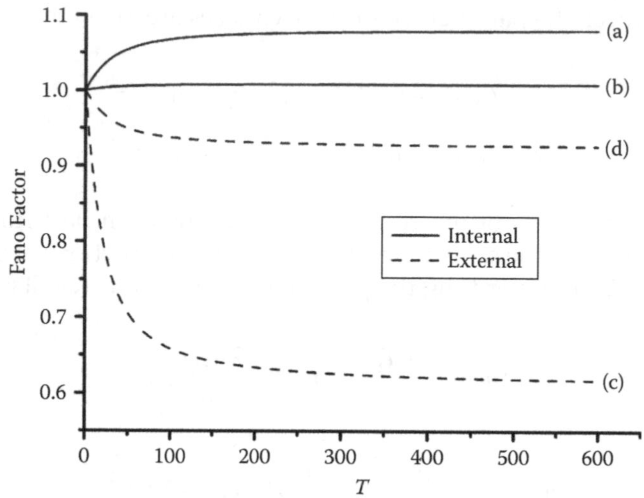

FIGURE 5.4
Fano factor for the monitored, limited birth–death process with $\mu = 0.005$ and $\lambda = 0.05$: (a) internal monitoring, $\eta = 0.025$; (b) internal monitoring, $\eta = 0.0025$; (c) external monitoring, $\eta = 0.025$; (d) external monitoring, $\eta = 0.0025$. Compare with Figure 4.6 (see Chapter 4). Note that the Fano factor is independent of the upper limit U.

This means that the externally monitored counts have a tendency to be separated in time, a property that is highlighted by the distribution of the time between events that can be calculated from result (5.18b) using the relation (2.54) (see Chapter 2). The simplest case is $U = 1$ when the following formula is obtained

$$w_1(t) = \frac{1-x^2}{x}\exp(-\gamma_e)\sinh(\gamma_e x), \tag{5.25}$$

where $x^2 = 1 - 4\eta\lambda/(\mu+\lambda+\eta)^2$. This quantity is plotted in Figure 5.5 and may be contrasted with the monotonically decreasing result obtained for the Poisson case given by Equation (2.57) (see Chapter 2). The tendency for events to be separated in time is a potentially useful property that can be exploited in data modelling. In quantum optics such a series of photon detections is said to be *anti-bunched*. Note that relation (5.24) reduces to the second factorial moment rather than the normalised mean square as $\tau, T \to 0$ ($\tau \geq T$, see 5.22) and does not contravene the Schwarz inequality.

We conclude that the limited birth–death process leads to a population that exhibits sub-Poisson fluctuations and that an anti-bunched series of events is generated by individuals leaving it in proportion to the number present, that is, through external monitoring.

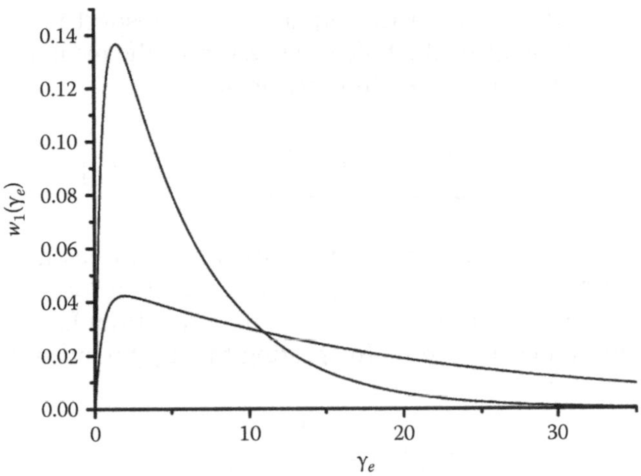

FIGURE 5.5
The distribution of interevent times for the monitored limited birth–death process, plotted as a function of the parameter γ_e. Here, $\mu = \lambda = 1$. The curve with the higher peak is for $\eta = 0.5$ and the other for $\eta = 0.1$.

5.4 A Birth-Limited Death–Immigration Process

As we have already remarked, in Chapter 3 we found that a simple birth–death process leads to a population that either grows without limit or becomes extinct according to whether the birth rate is larger than or smaller than the death rate. In the last section we saw how this process could be stabilised by limiting the births. If the death rate is greater than the birth rate, it is plausible that extinction might be avoided by limiting the number of deaths in some way.

Suppose that the deaths in a birth–death process are proportional to the difference between the number of individuals present and a fixed lower bound L so that there is no possibility of finding fewer than L individuals in the population, that is, $P_N = 0$ for $N < L$. This is akin to 'cover' in a predator–prey model that effectively hides a fraction of the prey from the predator species and thus prevents total annihilation of the prey population. The rate equation governing the evolution of such a population is

$$\frac{dP_N}{dt} = \mu(N+1-L)P_{N+1} - \mu(N-L)P_N + \lambda NP_N + \lambda(N-1)P_{N-1}. \qquad (5.26)$$

The equation for the corresponding generating function obtained by multiplying Equation (5.26) by $(1-s)^N$ and summing over N is

$$\frac{\partial Q}{\partial t} = -\mu s\frac{\partial Q}{\partial s} - \mu\frac{Ls}{1-s}Q + \lambda s(1-s)\frac{\partial Q}{\partial s}. \qquad (5.27)$$

Evaluating the derivative of this equation with respect to s at $s = 0$ and integrating over time gives the following result for the mean number in the population at time t when M are initially present:

$$\langle N(t)\rangle = \frac{\mu L}{\mu-\lambda} + \left(M - \frac{\mu L}{\mu-\lambda}\right)\exp[-(\mu-\lambda)t]. \tag{5.28}$$

Thus the population does not become extinct and does not grow without limit provided that $\mu > \lambda$. The generating function for the equilibrium population distribution in this case can be obtained from Equation (5.27) by setting the left-hand side to zero. Integrating the right-hand side then gives

$$Q_\infty(s) = \left(\frac{1-s}{1+\lambda s/(\mu-\lambda)}\right)^L. \tag{5.29}$$

This is a product of the generating function for a number distribution (i.e., one describing a fixed number, L, of individuals) and that of a negative binomial distribution. Following the discussion in Chapter 2, Section 2.3 consider the convolution

$$P_N = \sum_{R=0}^{N} P_R^{(1)} P_{N-R}^{(2)}. \tag{5.30}$$

Defining the generating function for P_N in the usual way, it is not difficult to show that it is simply the product of the generating functions for the convolved distributions:

$$Q(s) = Q^{(1)}(s)Q^{(2)}(s). \tag{5.31}$$

According to Equation (5.29), we can take $P_N^{(1)} = \delta_{NL}$ whilst $P_N^{(2)}$ is the negative binomial distribution (4.22) (see Chapter 4) with $\nu/\lambda = L$. Thus we find, $(\bar{N} = \mu L/(\mu-\lambda))$,

$$\begin{aligned} P_N &= \binom{N-1}{N-L}\frac{(\bar{N}/L)^{N-L}}{(1+\bar{N}/L)^N} \quad \textit{if } N \geq L \\ &= 0 \qquad\qquad\qquad\qquad\quad \textit{otherwise.} \end{aligned} \tag{5.32}$$

This shifted negative binomial distribution is plotted in Figure 5.6. As assumed initially, there is no possibility of finding fewer than L individuals in the population.

It is not difficult to generalise the model to include immigration. This simply requires the addition of the term $-\nu s Q$ to the right-hand side of Equation

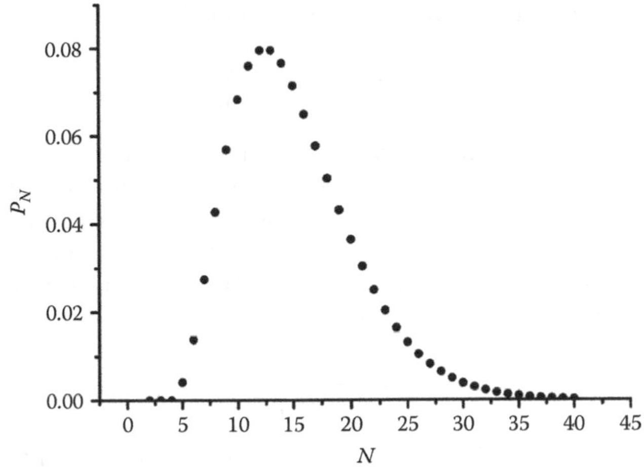

FIGURE 5.6
The shifted negative binomial distribution of Equation (5.32), with an average of 10 individuals in the population, and $L = 5$.

(5.27) and provided that $\mu > \lambda$ the long-term average number in the population then becomes

$$\bar{N} = \frac{\nu + \mu L}{\mu - \lambda}. \tag{5.33}$$

When $\nu = 0$ this reduces to the long-time limit of Equation (5.28) and when $L = 0$ it reduces to the usual result for a birth death immigration process (4.16) (see Chapter 4). The generating function for the equilibrium distribution when immigration is included in the model is given by

$$Q_\infty(s) = \frac{(1-s)^L}{(1+\lambda s/(\mu-\lambda))^{L+\nu/\lambda}}. \tag{5.34}$$

This again generates the convolution of a number distribution with a negative binomial distribution and reduces to the expected results when there is no immigration or no limit L on deaths.

The second factorial moment of the population number fluctuations can easily be obtained from Equation (5.34) and after normalisation may be written

$$\frac{\langle N(N-1)\rangle}{\langle N\rangle^2} = 1 + \frac{\mu L(2\lambda - \mu) + \nu\lambda}{(\mu L + \nu)^2}. \tag{5.35}$$

It is clear that this quantity can be greater than or less than unity so that the population distribution may be sub- or super-Poisson according to the values taken by the transition rates.

The Fano factor corresponding to Equation (5.35) is

$$F = 1 + \frac{\lambda}{\mu - \lambda} - \frac{\mu L}{\nu + \mu L}. \tag{5.36}$$

In the absence of immigration this implies super-Poisson statistics when $\lambda < \mu < 2\lambda$ and sub-Poisson statistics in the region $\mu > 2\lambda$. Note that results (5.33) to (5.36) also apply for the case $\lambda = 0$, that is, to a limited-death immigration model. In this case Equation (5.34) reduces to the convolution of a number distribution and a Poisson distribution, and the model predicts only sub-Poisson statistics with the Fano factor

$$F = 1 - \frac{\mu L}{\nu + \mu L}. \tag{5.37}$$

This reduces to the result $F = 1$ for a Poisson distribution when $L = 0$ and when $\mu L >> \nu$ approaches the result $F = 0$ for a distribution characterising a fixed number of individuals.

The full time-dependent generating function for the birth–limited death–immigration problem can be obtained by writing the solution of Equation (5.27) in the form

$$Q(s;t) = (1-s)^L \hat{Q}(s;t). \tag{5.38}$$

This transforms the equation into the form

$$\frac{\partial \hat{Q}}{\partial t} = -\mu s \frac{\partial \hat{Q}}{\partial s} + \lambda s(1-s)\frac{\partial \hat{Q}}{\partial s} - (\nu + \lambda L)s\hat{Q}. \tag{5.39}$$

Equation (5.39) is seen to have a structure that is identical with Equation (4.11) for the birth–death–immigration process and the solution of Equation (5.39) can be obtained from Equation (4.11) by making the parameter transformation $\nu \to \nu + \lambda L$. Thus we can use the time-dependent solution (4.34) of the birth–death–immigration process to obtain a result for the limited death case, remembering that in this case if the initial number in the population is M then $\hat{Q}(s;0) = (1-s)^{M-L}$.

$$Q(s,t) = (1-s)^L \left(1 + \frac{\lambda s}{\mu - \lambda}[1-\theta]\right)^{-(L+\nu/\lambda)} \left(1 - \frac{s\theta}{1 + \lambda s(1-\theta)/(\mu - \lambda)}\right)^{M-L}. \tag{5.40}$$

Here $\theta = \exp[-(\mu - \lambda)t]$. All the higher-order joint statistics can be obtained from this conditional generating function. In particular the bilinear moment can be calculated using formula (5.11):

$$\frac{\langle N(0)M(t)\rangle}{\langle N\rangle^2} = 1 + \frac{\mu(\nu + \lambda L)}{(\nu + \mu L)^2}\exp\left[-(\mu - \lambda)t\right].$$

This reduces to the correct value for the normalised mean square of the number of individuals in the population at $t = 0$ and to unity at long times when the fluctuations become uncorrelated. In general it exhibits the positive correlation between the numbers of individuals in the population at two different times expected for a first-order Markov process.

5.5 The Effect of Monitoring a Birth–Limited Death Process

Following the pattern of previous treatments of the problem, it is not difficult to write equations for the evolution of the probability distribution of internally and externally monitored counts for the birth–limited death process. For the internal case we obtain

$$\frac{dp_{N,n}}{dT} = \mu(N+1-L)p_{N+1,n} - \mu(N-L)p_{N,n} + \lambda N p_{N,n} + \lambda(N-1)p_{N-1,n} + \\ + \eta(N-L)(p_{N,n-1} - p_{N,n}), \tag{5.41}$$

whilst for the external measurement

$$\frac{dp_{N,n}}{dT} = \mu(N+1-L)p_{N+1,n} - \mu(N-L)p_{N,n} + \lambda N p_{N,n} + \lambda(N-1)p_{N-1,n} + \\ + \eta(N+1-L)p_{N+1,n-1} - \eta(N-L)p_{N,n}. \tag{5.42}$$

Note that we have assumed that the additional counting terms in these equations are proportional to the excess of individuals above the lower bound L. This restriction is essential in the case of external monitoring because a normal 'emigration' term, proportional to the number of individuals present, would allow the population number to access values that were less than the lower bound L and be contrary to the spirit of the model. The evolution of the corresponding generating functions is governed by the following equations for the internal and external monitoring cases, respectively:

$$\frac{\partial q_i}{\partial T} = -\mu s \frac{\partial q_i}{\partial s} - \mu \frac{Ls}{1-s} q_i + \lambda s(1-s)\frac{\partial q_i}{\partial s} + \eta z(1-s)\frac{\partial q_i}{\partial s} + \eta z L q_i, \tag{5.43}$$

$$\frac{\partial q_e}{\partial T} = -\mu s \frac{\partial q_e}{\partial s} - \mu \frac{Ls}{1-s} q_e + \lambda s(1-s)\frac{\partial q_e}{\partial s} + \eta(z-s)\frac{\partial q_e}{\partial s} + \eta \frac{(z-s)}{1-s} L q_e. \tag{5.44}$$

Both these equations are reduced by writing the generating functions in the form

$$q(s,z;T) = (1-s)^L \hat{q}(s,z;T) \tag{5.45}$$

This leads to

$$\frac{\partial \hat{q}_i}{\partial T} = -\left[\lambda s^2 + s(\mu - \lambda + \eta z) - \eta z\right]\frac{\partial \hat{q}_i}{\partial s} - \lambda L s \hat{q}_i, \tag{5.46}$$

$$\frac{\partial \hat{q}_e}{\partial T} = -\left[\lambda s^2 + s(\mu - \lambda + \mu) - \eta z\right]\frac{\partial \hat{q}_e}{\partial s} - \lambda L s \hat{q}_e. \tag{5.47}$$

These equations for $\hat{q}$ are of exactly the same form as those for the generating functions (4.55) and (4.56) characterising the conventional birth death immigration process (see Chapter 4). Indeed, immigration can again be included in the present problem by simply making the transformation $\lambda L \to \nu + \lambda L$ in Equations (5.46) and (5.47). The counting distributions are obtained from the generating functions evaluated at $s = 0$:

$$q(0,z;T) = \hat{q}(0,z;T). \tag{5.48}$$

Thus the final results for the generating functions of the counting distributions (if immigration is included) follow directly from the solutions (4.61) and (4.65) (see Chapter 4):

$$q_i(0,z;T) = \frac{\exp[(L+\alpha)\gamma_i]}{\left[\cosh y_i + (1+(y_i-\gamma_i)^2/2y_i\gamma_0)\sinh y_i\right]^{L+\alpha}} \tag{5.49}$$

$$q_e(0,z;T) = \frac{\exp[(L+\alpha)\gamma_e]}{[\cosh y_e + (\gamma_e/2y_e + y_e/2\gamma_e)\sinh y_e]^{L+\alpha}}. \tag{5.50}$$

The parameters in these formulae are the same as those defined in Equation (4.62) (Chapter 4):

$$\begin{aligned} &\alpha = \nu/\lambda; \quad \gamma_i = (\mu + \eta z - \lambda)T/2; \quad y_i^2 = \gamma_i^2 + \eta\lambda T^2 z \\ &\gamma_0 = \gamma_i(z=0); \quad \gamma_e = (\mu + \eta - \lambda)T/2; \quad y_e^2 = \gamma_e^2 + \eta\lambda T^2 z. \end{aligned} \tag{5.51}$$

All the single interval moments of the counting distributions for internal and external monitoring can be calculated in principle from Equations (5.49) and (5.50). The lower moments can be deduced from Equations (4.63) to (4.65):

$$\langle n_i \rangle = \eta T \frac{\nu + \lambda L}{\mu - \lambda}, \quad \langle n_e \rangle = \eta T \frac{\nu + \lambda L}{\mu + \eta - \lambda}, \tag{5.52}$$

$$n_i^{[2]} = 1 + \frac{\mu}{(\nu + \lambda L)\gamma_0}\left\{1 - \frac{1}{2\gamma_0}\left[1 - \exp(-2\gamma_0)\right]\right\} \tag{5.53}$$

$$n_e^{[2]} = 1 + \frac{\lambda}{(\nu + \lambda L)\gamma_e}\left\{1 - \frac{1}{2\gamma_e}[1 - \exp(-2\gamma_e)]\right\}. \tag{5.54}$$

Evidently the counting distributions are identical to those of the birth–death–immigration process studied in Chapter 4 but with an enhanced immigration rate $\nu + \lambda L$. The counting distributions are therefore super-Poisson and the interesting sub-Poisson regime of the population number fluctuations is not reflected in the counting statistics based on Equations (5.41) and (5.42). Moreover, the correlation function of the internally and externally monitored counts can be obtained from Equations (4.80) and (4.81) by similarly making the transformation $\nu \to \nu + \lambda L$ so that the sequence of counting events are bunched. These results reflect the fact that the shifted character of the population number distribution that gives rise to the 'non-classical' behaviour noted in the last section is removed by the measuring process defined in Equations (5.41) or (5.42).

In the case of internal monitoring the count rate is not constrained by the need to preserve the basic features of the model since the counting process does not affect the population evolution. Thus we could adopt the model used in all the earlier examples where it was assumed that the count rate was simply proportional to the number present in the population:

$$\frac{dp_{N,n}}{dT} = \mu(N+1-L)p_{N+1,n} - \mu(N-L)p_{N,n} + \lambda N p_{N,n} + \lambda(N-1)p_{N-1,n} + \\ + \eta N(p_{N,n-1} - p_{N,n}). \tag{5.55}$$

The equivalent generating function equation now becomes

$$\frac{\partial q_i}{\partial T} = -\mu s\frac{\partial q_i}{\partial s} - \mu\frac{Ls}{1-s}q_i + \lambda s(1-s)\frac{\partial q_i}{\partial s} + \eta z(1-s)\frac{\partial q_i}{\partial s}. \tag{5.56}$$

The transformation (5.45) does not reduce this to an equation that has been solved previously, but it is not difficult to calculate the second moment from (5.56):

$$n_i^{[2]} = 1 + \frac{\lambda}{\mu L\gamma_0}\left\{1 - \frac{1}{2\gamma_0}[1-\exp(-2\gamma_0)]\right\}. \tag{5.57}$$

Here γ_0 is given by Equation (5.51) as before. In the short integration time limit (5.57) reduces to $\langle N^2\rangle/\langle N\rangle^2$ as found for all previous models in the case of internal monitoring. Again, the sub-Poisson regime exhibited by the basic population model is not reflected in the measured count rate.

We conclude that the birth–limited death–immigration process leads to a population that exhibits various regimes of super- and sub-Poisson fluctuation behaviour but that the simplest allowable monitoring schemes will not generate counts that are sub-Poisson or anti-bunched for any values of the model parameters.

5.6 Odd–Even Effects in a Population Driven by Immigrant Pairs

We now consider the evolution of a population in which deaths occur singly as in our previous models but into which immigration takes place in the form of pairs of individuals rather than singly as in the death–immigration process discussed in Chapter 3. This is another example of a model that exhibits 'non-classical' behaviour and it also provides the basis for the more general, multiple-immigrant models developed in Chapter 6. The population transitions for this process are illustrated in the level diagram shown in Figure 5.7 and the evolution of the population is described by the rate equation

$$\frac{dP_N}{dt} = \mu(N+1)P_{N+1} - \mu N P_N - \nu_2 P_N + \nu_2 P_{N-2}. \tag{5.58}$$

The novel element of the process lies in the final term that expresses an increase in a population of N individuals due to pairs of immigrants entering a population of $N - 2$ at a rate ν_2. The equivalent generating function equation obtained by multiplying Equation (5.58) by $(1 - s)^N$ and summing over N takes the form

$$\frac{\partial Q}{\partial t} = -\mu s \frac{\partial Q}{\partial s} + \nu_2[(1-s)^2 - 1]Q. \tag{5.59}$$

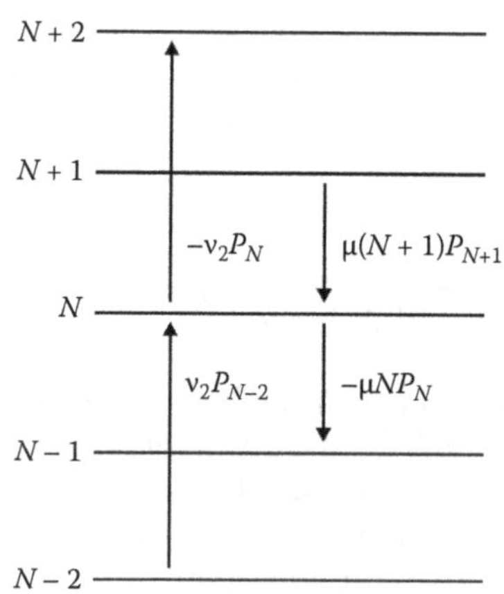

FIGURE 5.7
Transition diagram for a population with immigrations and deaths, in which the immigrations consist of pairs of individuals.

Using the method of Laplace transforms described in Chapter 4, Section 4.4, this equation can be solved exactly to give the time evolution of the generating function:

$$Q(s;t) = Q(se^{-\mu t};0)\exp\left\{\frac{\nu_2}{2\mu}\left[se^{-\mu t}(4 - se^{-\mu t}) - s(4-s)\right]\right\}. \tag{5.60}$$

At long times this approaches the equilibrium solution

$$Q_\infty(s) = \exp\left\{-\frac{\nu_2 s}{2\mu}(4-s)\right\}. \tag{5.61}$$

The probabilities and their moments can be obtained by evaluating the derivatives of this result with respect to s at $s = 0$; for example, the mean and normalised second factorial moments of the distribution are

$$\bar{N} = \frac{2\nu_2}{\mu}, \tag{5.62}$$

$$N^{[2]} = 1 + \frac{\mu}{4\nu_2}. \tag{5.63}$$

For higher moments and distributions, however, this process is expedited by recalling that the generating function for the Hermite polynomials $H_r(x)$ has the form

$$\exp(2xz - z^2) = \sum_{r=0}^{\infty} H_r(x)\frac{z^r}{r!}. \tag{5.64}$$

It is not difficult to show from this formula that the equilibrium generating function (5.61) can be expanded in the following two ways

$$Q_\infty(s) = \sum_{r=0}^{\infty} H_r\left(\sqrt{-\bar{N}}\right)\frac{\left(-\bar{N}/4\right)^{r/2} s^r}{r!}, \tag{5.65}$$

$$Q_\infty(s) = \exp\left(-3\bar{N}/4\right)\sum_{r=0}^{\infty} H_r(-\sqrt{\bar{N}}/2)\frac{(-\bar{N}/4)^{r/2}(1-s)^r}{r!}. \tag{5.66}$$

Since the Hermite polynomials of imaginary argument may be expressed in terms of Laguerre polynomials, the equilibrium moments and distribution of the immigrant pair–death process can now be written from these two expansions in the following form:

$$N^{[2r]} = \frac{r!L_r^{-1/2}(-\bar{N})}{\bar{N}^r}; \quad N^{[2r+1]} = \frac{r!L_r^{1/2}(-\bar{N})}{\bar{N}^r}, \tag{5.67}$$

$$\begin{aligned} P_{2N} &= \frac{N!\bar{N}^N}{(2N)!}\exp\left(-\frac{3\bar{N}}{4}\right)L_N^{-1/2}\left(-\frac{\bar{N}}{4}\right); \\ P_{2N+1} &= \frac{N!\bar{N}^{N+1}}{2(2N+1)!}\exp\left(-\frac{3\bar{N}}{4}\right)L_N^{1/2}\left(-\frac{\bar{N}}{4}\right). \end{aligned} \tag{5.68}$$

These results exhibit mathematically the odd–even asymmetry that is an essential characteristic of the model. However, the effect, though present, is not particularly marked in plots of the distribution (5.68) shown in Figure 5.8. According to Equations (5.62) and (5.63) the fluctuations are super-Poisson with a Fano factor

$$F = \frac{3}{2}. \tag{5.69}$$

There is no continuous analogue of the process, however, since even and odd has no meaning for a continuous variable.

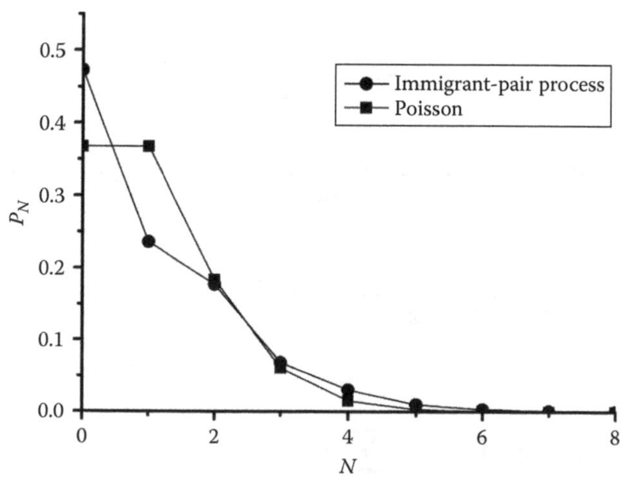

FIGURE 5.8
Distribution for the immigrant pair process, with a mean of 1 individual, Equation (5.68). A Poisson distribution of the same mean is shown for comparison.

The number correlation function or bilinear moment for the process can be obtained from Equation (5.58) by methods described in Chapter 3, Section 3.3 or directly from the solution (5.60) using the definition

$$\langle N(0)M(t)\rangle = \sum_{M,N=0}^{\infty} NMP_N P(M;t|N;0) = -\sum_{N=0}^{\infty} NP_N \left.\frac{\partial Q(s;t|N;0)}{\partial s}\right|_{s=0}. \tag{5.70}$$

In this formula the transient solution (5.60) must be evaluated subject to the condition that exactly N individuals were present initially, that is, $Q(s;0) = (1-s)^N$. A simple calculation gives for the normalised correlation function

$$G(t) = 1 + \frac{3}{2\bar{N}}\exp(-\mu t). \tag{5.71}$$

This shows the usual behaviour expected of a first-order Markov process, with exponential decay from the value $\langle N^2\rangle/\bar{N}^2$ at $t = 0$ to unity at times that are large compared with the fluctuation time of the process, μ^{-1}.

5.7 Monitoring an Immigrant Pair–Death Process

We can calculate the statistical properties of counts obtained through internal or external monitoring by adding extra terms to the rate Equation (5.58) as in previous calculations. Thus for internal monitoring we obtain

$$\begin{aligned}\frac{dP_{N,n}}{dT} &= \mu(N+1)P_{N+1,n} - \mu NP_{N,n} + \nu_2 P_{N-2,n} - \nu_2 P_{N,n}\\ &\quad + \eta NP_{N,n-1} - \eta NP_{N,n},\end{aligned} \tag{5.72}$$

whilst for external monitoring the following rate equation applies

$$\begin{aligned}\frac{dP_{N,n}}{dT} &= \mu(N+1)P_{N+1,n} - \mu NP_{N,n} + \nu_2 P_{N-2,n} - \nu_2 P_{N,n}\\ &\quad + \eta(N+1)P_{N+1,n-1} - \eta NP_{N,n}.\end{aligned} \tag{5.73}$$

The equivalent equations for the internal and external counting generating functions are

$$\frac{\partial q_i}{\partial T} = [\eta z - s(\mu + \eta z)]\frac{\partial q_i}{\partial s} + \nu_2[(1-s)^2 - 1]q_i, \tag{5.74}$$

$$\frac{\partial q_e}{\partial T} = -[\mu s + \eta(s - z)]\frac{\partial q_e}{\partial s} + \nu_2[(1-s)^2 - 1]q_e. \tag{5.75}$$

Both of these equations can be solved exactly by the method of Laplace transforms described in Chapter 4, treating z as an extra parameter. However, the solution for internal monitoring is a relatively complicated function of this parameter and for this case the low moments are most easily obtained by taking derivatives of Equation (5.74) with respect to z and s at $s = 0$. Thus we obtain

$$\frac{d\langle n_i\rangle}{dT} = \eta\bar{N}; \quad \frac{d\langle n_i(n_i-1)\rangle}{dT} = 2\eta\langle n_i N\rangle$$
$$\frac{d\langle n_i N\rangle}{dT} + \mu\langle n_i N\rangle = \eta\langle N^2\rangle + 2\nu_2\langle n_i\rangle. \tag{5.76}$$

These equations may be solved with the help of results (5.62) and (5.63) to give

$$\langle n_i\rangle = \frac{2\eta\nu_2 T}{\mu}, \tag{5.77}$$

$$n_i^{[2]} = 1 + \frac{3}{2\nu_2 T}\left\{1 - \frac{1}{\mu T}[1-\exp(-\mu T)]\right\}. \tag{5.78}$$

As we have come to expect from previous calculations, in the short integration time limit the normalised second factorial moment of the internally monitored counting distribution reduces to the normalised *mean square* of the population number.

The generating function for the externally monitored counting distribution can be written exactly in the compact form

$$q_e(z) = \exp\left[-\frac{\langle n_e\rangle z}{2}(2 - \quad (T)z)\right], \tag{5.79}$$

where

$$\langle n_e\rangle = \frac{2\eta\nu_2 T}{\mu+\eta}$$
$$(T) = \frac{\eta}{\mu+\eta}\left[1 - \frac{1-\exp[-2\gamma_2]}{2\gamma_2}\right] \tag{5.80}$$
$$\gamma_2 = (\mu+\eta)T/2.$$

The normalised second factorial moment can easily be calculated to be

$$n_e^{[2]} = 1 + \frac{1}{\langle n_e\rangle}\quad(T). \tag{5.81}$$

This reduces to the normalised second factorial moment of the population itself, Equation (5.63), in the small integration time limit, assuming the death rate parameter is increased by η. In the long integration time limit we find from Equations (5.78) and (5.81)

$$\lim_{\mu T\to\infty} F_i = 1 + \frac{3\eta}{\mu}, \tag{5.82}$$

$$\lim_{(\mu+\eta)T\to\infty} F_e = 1 + \frac{\eta}{\mu+\eta}. \tag{5.83}$$

Although these results together with the moments (5.78) and (5.81) show that the population is super-Poisson they give no indication of odd–even effects. However, in the case of an externally counted population it is not difficult to obtain expressions for the full distribution of counts. To see this we first note that result (5.79) is a similar function of z to (5.61) for the population itself and to the generating function for the Hermite polynomials. This can be exploited as in the last section to obtain closed forms for the moments and distribution of the externally monitored counts in terms of Laguerre polynomials:

$$n_e^{[2r]} = r!L_r^{-\frac{1}{2}}\left(-\frac{\langle n_e\rangle}{2}\right)\left(\frac{2}{\langle n_e\rangle}\right)^r; \quad n_e^{[2r+1]} = r!L_r^{\frac{1}{2}}\left(-\frac{\langle n_e\rangle}{2}\right)\left(\frac{2}{\langle n_e\rangle}\right)^r, \tag{5.84}$$

$$p_{2n} = \frac{n!}{(2n)!}L_n^{-\frac{1}{2}}\left(-\frac{\langle n_e\rangle(1-\ \ )^2}{2}\right)(2\langle n_e\rangle\ \ )^n \exp[-\langle n_e\rangle(1-\ \ /2)], \tag{5.85}$$

$$p_{2n+1} = \frac{\langle n_e\rangle(1-\ \ )n!}{(2n+1)!}L_n^{\frac{1}{2}}\left(-\frac{\langle n_e\rangle(1-\ \ )^2}{2}\right)(2\langle n_e\rangle\ \ )^n \exp[-\langle n_e\rangle(1-\ \ /2)]. \tag{5.86}$$

The distributions plotted from these results are shown in Figure 5.9 and exhibit striking odd–even effects for some parameter values, in contrast to Figure 5.8 where they were hardly noticeable in the population distribution itself.

Note that in the short integration time limit or when $\mu >> \eta$ so that $\varsigma \approx 0$, Equation (5.79) reduces to the generating function for a Poisson distributed number of *individuals*, whereas in the case when $\varsigma \approx 1$ a Poisson distributed number of *pairs* is predicted. For this to occur, according to Equation (5.86), ς must be closer to unity if $\langle n_e\rangle$ is large. Although $\langle n_e\rangle$ and ς can be treated independently, restricting the value of ς has implications for the population mean $\bar{N}$ since $\langle n_e\rangle = 2(\eta/(\mu+\eta))\bar{N}\gamma_e$. In order to achieve values of ς close to unity it is necessary for $\mu << \eta$ and $\gamma_e << 1$ so the optimum situation for observing odd–even effects corresponds to $\bar{N} << 1$. In other words the largest effects occur when the population mean is small and *all* emigrants from the population are counted for periods long compared with the characteristic fluctuation time.

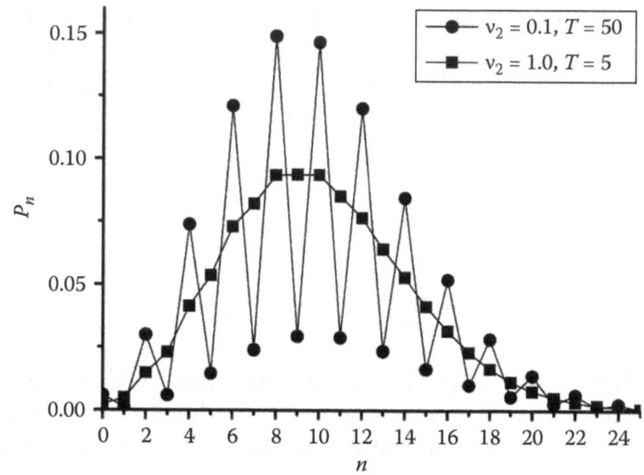

FIGURE 5.9
External monitoring distribution for the immigrant pair process, showing odd–even effects. Here, $\eta = 1$ and $\mu = 0$, so there is no conventional death element. Both plots are for $\langle n_e \rangle = 10$.

We have used the qualification 'all' here advisedly because in the case of external monitoring, we have hitherto ignored the possibility that *not all individuals that leave the population are counted.* In Chapter 2 we discussed the process of Bernoulli sampling by which a train of events is counted by random selection, and showed that this can be taken into account by multiplying z in the generating function $q_e(z;T)$ for the counting distribution by an additional *efficiency* factor, ξ. For all the population models that we have considered so far this scales the mean count rate and does not change the higher *normalised* factorial moments of the counting distribution. For example, in the present model $n_e^{[2r]}$ is dependent only on the ratio of $\langle n_e \rangle$ and ς, which according to Equation (5.79) will both be scaled by any efficiency factor ξ. However, $\langle n_e \rangle$ and ς appear separately in formulae (5.85) and (5.86) for the distribution, and the effect of inefficient counting is to reduce the size of the even–odd effects visible in Figure 5.9 by effectively adding single-individual randomness through the counting process.

The correlation functions of the number of counted individuals can be found using the method described in Chapter 4, Section 4.7; for example, for the case of external monitoring we obtain

$$g_e(\tau) = 1 + \frac{\mu}{4\nu_2}\left(\frac{\sinh\gamma_2}{\gamma_2}\right)^2 \exp[-(\mu+\eta)\tau]. \tag{5.87}$$

The parameter γ_2 in this equation is defined in Equation (5.80). As expected, this shows the usual Markovian exponential decay, with the fluctuation magnitude reduced by the integration process.

5.8 Summary

- In this chapter we have studied three discrete population models that have no continuous counterparts.
- We have shown how limiting the births or deaths in a simple birth–death model can lead to a population with equilibrium fluctuations that are sub-Poisson.
- For these processes we have calculated the integrated statistics for both internal and external monitoring schemes and shown that individuals leaving a population in which births are limited are anti-bunched with a preferred time between departures.
- We have also solved the rate equation for a population subject to immigrant pairs of individuals and shown that the equilibrium distribution exhibits odd–even effects.
- These effects are particularly marked in the distribution of individuals leaving such a population via the process of external monitoring, but may be masked if not all the individuals leaving the population are counted.

Problems

5.1 A population with exactly $M \leq K$ individuals present at $t = 0$ is governed by a limited birth–death process with $P_N = 0$ for $N > K$. Obtain expressions for the mean and normalised second factorial moment of the distribution of individuals present at time $t = \tau \geq 0$.

5.2 Show that external monitoring of the number of individuals leaving a population governed by a limited birth–death process leads to a Fano factor that lies between ½ and 1. How does this change if ξ, the efficiency of the counter, is less than unity?

5.3 Find a generating function for the equilibrium distribution of the number of individuals in a population in which both births and deaths are limited by the mechanisms defined in Sections 5.2 and 5.4. Show that the fluctuations are always sub-Poisson.

5.4 Find expressions for the mean, second normalised factorial moment and Fano factor of an immigrant pair–death process at time t given that there are no individuals present initially. Given that the pair immigration rate and the death rate are equal to 1 sec^{-1}, how long does it take for the variance of the population distribution to reach half its long time asymptotic value?

5.5 Obtain expressions in terms of elementary functions for the external counting probabilities p_0, p_1 and p_2 for an immigrant pair–death process. Find the condition that p_1 should be less than both p_0 and p_2.

Further Reading

E. Jakeman, 'Statistics of binomial number fluctuations,' *Journal of Physics A*, **23**, 2815–2825 (1990).

E. Jakeman, S. Phayre and E. Renshaw, 'The evolution and measurement of a population of pairs,' *Journal of Applied Probability*, **32**, 1048–1062 (1995).

E. Renshaw, *Stochastic Population Processes*, Oxford University Press, 2011, chap. 8.

6

The Death–Multiple Immigration Process

6.1 Introduction

This chapter introduces a flexible but rather powerful variation on the population processes that we have studied that can be used to generate a wide variety of models and the fluctuation phenomena that are associated with them. The immigrant pair model has deaths that occur by the usual mechanism and immigrants that arrive in couples. It is a straightforward generalisation to envisage a situation where the immigrants arrive in r-tuplets at rates ν_r that are particular to the order r. Provided that these rates are all positive quantities, they can be chosen arbitrarily. Indeed, the rates can be chosen so that the steady state distribution has a prescribed form. Consequently, once the techniques for deriving, solving and analysing the basic Markov population process have been mastered, the death–multiple immigration model can be used to generate the probability distributions and evolution characteristics of more complex systems that will be discussed in later chapters.

The death–multiple immigration model can also be designed to generate the same negative binomial population number distribution that was obtained in the case of the birth–death–immigration process described in Chapter 4. Attempts to distinguish the two models using their higher-order statistical properties provide insight into the more general problem of how the mechanisms underlying observed fluctuations in population numbers can be identified and what statistical measures can be used to test alternative theories.

6.2 Rate and Generating-Function Equations for the Process

The death–multiple immigration model is a first-order Markov process. The population randomly decreases through deaths that occur at a constant rate μ and in proportion to the instantaneous population size N. As we have already seen in Chapter 3, and with reference to Figure 6.1, the population 'level' N is affected by a death from $N + 1$ to N, and from N to $N - 1$. Increases

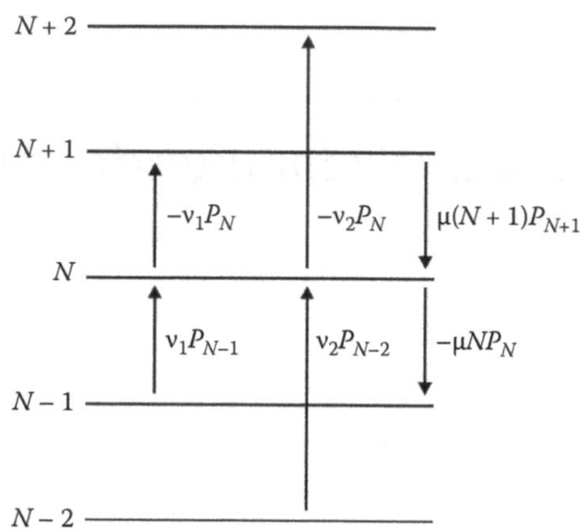

FIGURE 6.1
A death–multiple immigration process, with just the single and pair immigrations shown.

in the population happen by the immigration mechanism, which, unlike births, is independent of the existing population size. In the present model, immigrants can randomly enter the population singly, in pairs, or in multiples of size r. The singletons populate the level N via immigration from $N - 1$ and deplete it through a transition from N to $N + 1$. Pairs cause the level N to be populated via a transition from level $N-2$, but is once again depleted from level N. Continuing in this way, the maximum by which level N can be populated is by immigration of N individuals, that is, from level '0' to N, whereas level N can change to another higher level following an immigration of arbitrary order. If the rate at which r immigrants enter the population is ν_r, the rate equation for the death–multiple immigration process is then given by

$$\frac{dP_N}{dt} = \mu(N+1)P_{N+1} - \mu NP_N + (\nu_1 P_{N-1} - \nu_1 P_N) + (\nu_2 P_{N-2} - \nu_2 P_N) + \cdots$$
$$+ (\nu_N P_0 - \nu_N P_N) - \nu_{N+1}P_N - \cdots - \nu_{N+r}P_N - \cdots$$

which can be written succinctly as

$$\frac{dP_N}{dt} = \mu(N+1)P_{N+1} - \mu NP_N + \sum_{r=1}^{N} \nu_r P_{N-r} - P_N \sum_{r=1}^{\infty} \nu_r \tag{6.1}$$

where the first two terms encapsulate the death process and the latter two the multiple immigration process. The first summation is evidently a convolution-type term that couples the level N to $N - r$ via the simultaneous immigration of r individuals. The final term's summation to infinity allows immigrations of arbitrary order to occur. Note that setting the rates $\nu_1 = \nu$ and $\nu_r = 0$ for $r \geq 2$ obtains the death–immigration process of Chapter 3, and by suppressing the multiple immigrations of order $r = 1$ and $r \geq 3$ we obtain the pair process of the previous chapter.

The presence of the convolution-type term in Equation (6.1) implies that simple stationary solutions are not self-evident, especially since the process now involves the immigration rates of all orders. However, once the equivalent equation for the generating function is obtained, the mathematical structure of the death–multiple immigration process becomes more transparent, and as a consequence so too is the way that this structure can be exploited to an advantage.

The equation for the generating function can be obtained by multiplying both sides of Equation (6.1) by $(1 - s)^N$ and summing over N. The first three terms appearing in Equation (6.1) are familiar. Writing out the third term in full gives:

$$\sum_{N=0}^{\infty}\sum_{r=1}^{N}\nu_r P_{N-r}(1-s)^N = (1-s)\nu_1 P_0 + (1-s)^2\{\nu_1 P_1 + \nu_2 P_0\}$$

$$+(1-s)^3\{\nu_1 P_2 + \nu_2 P_1 + \nu_3 P_0\} + \dots$$

$$= \left\{(1-s)\nu_1 + (1-s)^2\nu_2 + \cdots + (1-s)^r\nu_r + \cdots\right\}P_0 +$$

$$\left\{(1-s)\nu_1 + (1-s)^2\nu_2 + \cdots + (1-s)^r\nu_r + \cdots\right\}(1-s)P_1 + \dots$$

$$\left\{(1-s)\nu_1 + (1-s)^2\nu_2 + \cdots + (1-s)^r\nu_r + \cdots\right\}(1-s)^m P_m + \dots$$

$$= \sum_{r=1}^{N}\nu_r(1-s)^r Q$$

and so Equation (6.1) becomes

$$\frac{\partial Q}{\partial t} = -\mu s\frac{\partial Q}{\partial s} + \left[\sum_{r=1}^{\infty}\nu_r\left\{(1-s)^r - 1\right\}\right]Q \tag{6.2}$$

with the first term on the right-hand side of the equality representing the deaths. The term involving the summation over the multiple immigration rates can be expressed as a function of s, so that Equation (6.2) can be written

$$\frac{\partial Q}{\partial t} = -\mu s\frac{\partial Q}{\partial s} + F(s)Q \tag{6.3}$$

and it is this form of the death–multiple immigration process that is the most suggestive in terms of its power and flexibility. Assuming that a stationary state can be attained, the left-hand side of Equation (6.3) vanishes and the resulting ordinary differential equation can be used in two ways. Should the multiple immigration rates and therefore the function F be prescribed, the ordinary differential equation for Q can be integrated to obtain the stationary-state generating function for the process,

$$Q(s,\infty) \equiv Q_\infty(s) = C\exp\left(\int_0^s ds' \frac{F(s')}{\mu s'}\right) \tag{6.4}$$

with C being selected to ensure that $Q(s = 0) = 1$. Alternatively, if a particular stationary-state generating function is required, the form that F must then adopt is given by

$$F(s) = \mu s \frac{d \ln Q_\infty(s)}{ds}. \tag{6.5}$$

In using Equation (6.5), it must be borne in mind that the form of F must ultimately be expressible in terms of the formula

$$F(s) = \sum_{r=0}^{\infty} \nu_r \{(1-s)^r - 1\} \tag{6.6}$$

with the constraint that the rates must *all* satisfy $\nu_r \geq 0$. Indeed, this constraint imparts a further interpretation on the form adopted by F. Suppose that ν_r are proportional to the probability that r-immigrants are introduced to the population, then it can be seen that the first term in the curly brackets is proportional to the generating function for this probability distribution of immigrants. If the constant of proportionality is ε, then the second term sums to exactly ε. Hence, it follows that F can be expressed in the form

$$F(s) = \varepsilon[\hat{Q}(s) - 1] \tag{6.7}$$

and the death–multiple immigration process can be interpreted as a population that is driven by an ensemble of fluctuating immigrant populations with an equilibrium generating function $\hat{Q}(s)$.

Some simple properties of the approach to equilibrium can be deduced without recourse to obtaining the full time-dependent solution for the generating function. Differentiating Equation (6.3) and setting $s = 0$ obtains the equation for the evolution of the mean size of the population:

$$\frac{\partial}{\partial t}\left(\frac{\partial Q}{\partial s}\right)\bigg|_{s=0} = -\mu\left(\frac{\partial Q}{\partial s} + s\frac{\partial^2 Q}{\partial s^2}\right)\bigg|_{s=0} + \left(\frac{dF}{ds}Q + F\frac{\partial Q}{\partial s}\right)\bigg|_{s=0}.$$

Now, from its definition, it is clear that $F(s = 0) = 0$, and $Q(s = 0,t) = 1$, hence

$$-\frac{d\langle N(t)\rangle}{dt} = \mu\langle N(t)\rangle + \left.\frac{dF}{ds}\right|_{s=0}$$

with solution

$$\langle N(t|M;0)\rangle = -\frac{1}{\mu}\left.\frac{dF}{ds}\right|_{s=0} + \left(M + \frac{1}{\mu}\left.\frac{dF}{ds}\right|_{s=0}\right)\exp(-\mu t) \tag{6.8}$$

where M is the population size initially. From Equation (6.5),

$$\frac{1}{\mu}\left.\frac{dF}{ds}\right|_{s=0} = \left.\left\{s\frac{\partial^2 Q_\infty(s)}{\partial s^2} + \frac{\partial Q_\infty(s)}{\partial s} - s\left(\frac{\partial Q_\infty(s)}{\partial s}\right)^2\right\}\right|_{s=0} = -\bar{N}, \tag{6.9}$$

where $\bar{N}$ is the mean of the population in its steady state. Consequently, the mean evolves according to

$$\langle N(t|M;0)\rangle = \bar{N} + (M - \bar{N})\exp(-\mu t).$$

This expression is identical with result (3.20) (see Chapter 3) for the death–single immigrant process and shows explicitly that the characteristic timescale on which the population changes is that associated with deaths, and indeed this is the only mechanism by which the population can decline.

To illustrate the use of Equation (6.5), we revisit the death–pair immigration model, for which the generating function is given by Equation (5.61) (Chapter 5)

$$Q_\infty(s) = \exp\left\{-\frac{\nu_2 s}{2\mu}(4 - s)\right\}, \tag{6.10}$$

which generates the Hermite polynomials. We see from Equation (6.5) that the function

$$F(s) = -2\nu_2 s + \nu_2 s^2 \tag{6.11}$$

is quadratic, which implies from Equation (6.6) that the immigrants involve pairs and possibly singletons, that is,

$$F(s) = -(\nu_1 + 2\nu_2)s + \nu_2 s^2. \tag{6.12}$$

Comparison with Equation (6.9) indicates that $\nu_1 = 0$ for consistency.

As another example, suppose that we require a model that, in equilibrium, leads to the Laguerre distributions defined in Chapter 2. These have more complicated generating functions, the simplest member of the class being given by

$$Q_\infty(s) = \frac{1}{1+N_2 s}\exp\left(-\frac{N_1 s}{1+N_2 s}\right). \tag{6.13}$$

Using Equation (6.5), this implies that

$$F(s) = -\mu s\left(\frac{N_1 + N_2(1+N_2 s)}{(1+N_2 s)^2}\right). \tag{6.14}$$

Expanding this expression in ascending powers of (1 – s) and comparing with Equation (6.6) obtains the immigration coefficients

$$\nu_r = \frac{\mu}{(1+N_2)^2}\left(\frac{N_2}{1+N_2}\right)^r\left(1+N_2+\frac{rN_1}{N_2}-N_1\right). \tag{6.15}$$

The $\{\nu_r\}$ are all positive and the process realisable only if the *single* immigrant coefficient ν_1 is positive, that is, if $N_1 \le 1 + N_2 + N_1/N_2$.

Result (6.15) highlights possible limitations of the model when attempting to generate a discrete Markov process with arbitrary equilibrium distribution. In some circumstances this type of problem can be overcome by including a simple birth process in Equation (6.1). The general time-dependent solution of this model is

$$Q(s;t|M;0) = [1 - f(s;t)]^M \exp[G(s) - G(f(s;t))] \tag{6.16}$$

where

$$G(x) = \int_0^x dy \frac{F(y)}{y(\mu - \lambda + \lambda y)}$$

$$f(s;t) = \frac{s\exp[-(\mu-\lambda)t]}{1+\dfrac{\lambda s}{\mu-\lambda}(1-\exp[-(\mu-\lambda)t])}.$$

The generating function (6.16) reduces to (6.4) in the long time limit upon setting the birth rate $\lambda = 0$.

6.3 Negative Binomial Fluctuations Revisited

Chapter 4 established that the birth–death–immigration process had negative binomial number fluctuations in its steady state with the generating function given by

$$Q(s) = \left(1 + \frac{\bar{N}}{\beta}s\right)^{-\beta}. \tag{6.17}$$

Using Equation (6.5) gives the function F as

$$F(s) = -\mu\bar{N}s\left(1 + \frac{\bar{N}}{\beta}s\right)^{-1} \tag{6.18}$$

which on writing $s = 1 - (1 - s)$, expanding F in powers of $(1 - s)$ and comparing the result with the expansion given by Equation (6.6) identifies the multiple immigration rates to be of the form $\nu_r = a\zeta^r$ where

$$a = \frac{\mu\beta}{\left(1 + \bar{N}/\beta\right)} \quad \text{and} \quad \zeta = \frac{\bar{N}/\beta}{\left(1 + \bar{N}/\beta\right)},$$

and it can be seen that the rates are all positive. In fact, the rates are the terms in a geometrical series, and the geometrical series is one member of the negative binomial class of distributions corresponding to the Bose–Einstein or thermal distribution. Therefore, with this choice of multiple immigration rates, the death–multiple immigration process generates the entire class of negative binomial distributions from a single member of that class.

Although the fluctuations in the steady state have the same distribution as that for the birth–death immigration process, the processes themselves differ significantly as can be seen from the simulations shown in Figure 6.2 and so we can expect that the routes to this stationary state should differ, and that measurements involving samples taken from the different processes taken at different times would also be dissimilar. The method for obtaining the time-dependent solution was demonstrated in Chapter 4, and it is given by

$$Q(s,t) = Q(s\theta(t),0)\left(\frac{1 + (s\bar{N}/\beta)\theta(t)}{1 + (s\bar{N}/\beta)}\right)^{\beta} \tag{6.19}$$

where $\beta = a\zeta/[\mu(1-\zeta)]$, and $\bar{N} = a\zeta/[\mu(1-\zeta)^2]$, with $\theta(t) = \exp(-\mu t)$. Note that when $t = 0$, $\theta(t) = 1$ and Equation (6.19) is an identity, whereas when $t \to \infty$,

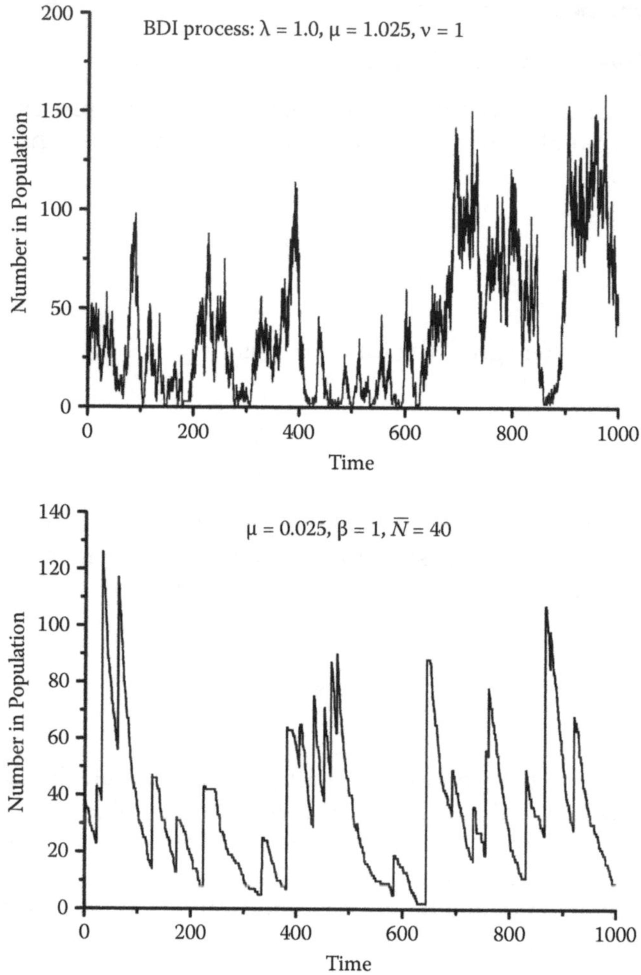

FIGURE 6.2
Birth–death–immigration process (upper) and death–multiple immigration process (lower), with the same distributions and correlation functions.

$\theta(t) \to 0$ and Equation (6.19) becomes Equation (6.17). Consequently, these rates produce negative-binomial number fluctuations at asymptotically large times whatever the initial state of the population.

Comparing Equation (6.19) with Equation (4.34) shows that the time dependence of the processes do indeed differ and so the routes to the stationary-state solution will be different. Moreover, it would be reasonable to expect that measurements made on the population at different times would yield different results. Figure 6.3 compares the evolution of the probability distribution from identical initial to identical final states

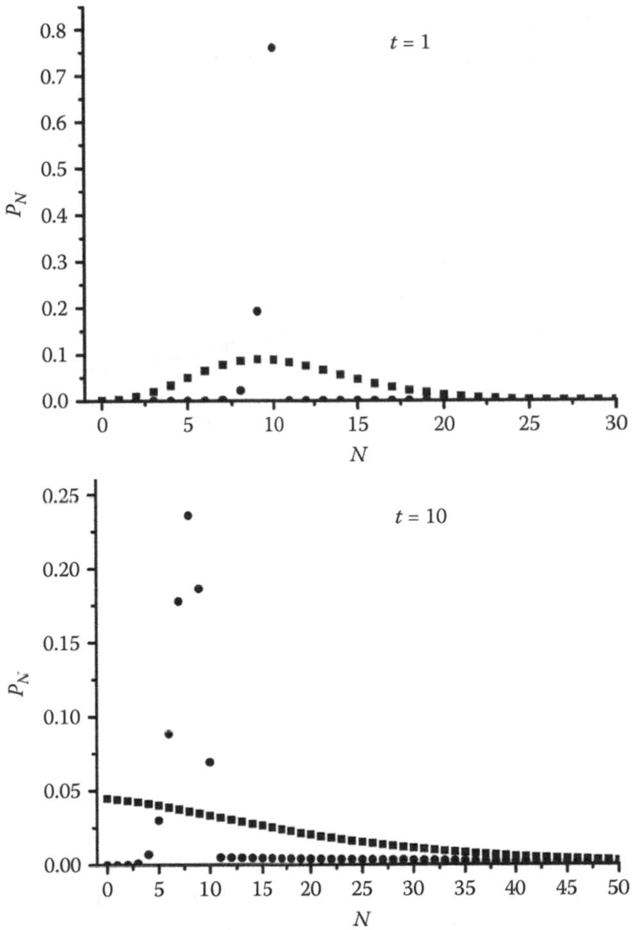

FIGURE 6.3
Evolution of the probability distribution with time, starting with $N = 10$ at $t = 0$. Birth–death–immigration process (squares) and death-multiple immigration process (circles), with the same parameters as in Figure 6.2. (*Continued*)

for the birth death immigration and death–multiple immigration processes. The evolution of the second-factorial moment can be deduced from Equation (6.19) by differentiating twice with respect to s and then by setting $s = 0$, which obtains

$$\langle N(t)(N(t)-1)\rangle = M(M-1)\theta^2 + 2M\bar{N}\theta(1-\theta) + \frac{\bar{N}^2}{\beta}(1-\theta)(2+(\beta-1)(1-\theta)) \tag{6.20}$$

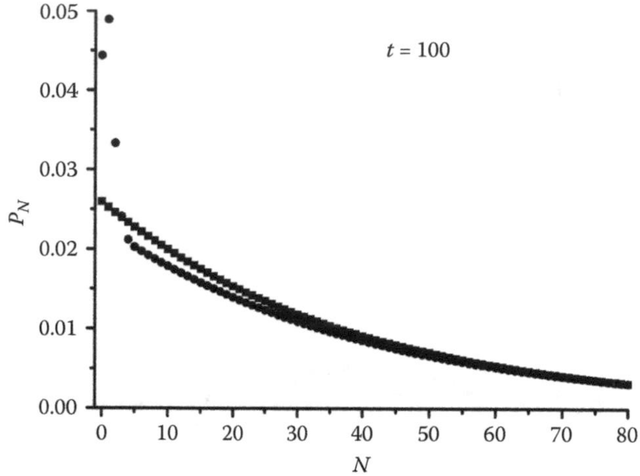

FIGURE 6.3 *(Continued)*
Evolution of the probability distribution with time, starting with $N = 10$ at $t = 0$. Birth–death–immigration process (squares) and death–multiple immigration process (circles), with the same parameters as in Figure 6.2.

and it can be seen that this reduces to $M(M-1)$ as $t \to 0$, and as $t \to \infty$ to the value

$$N^{[2]} \equiv \frac{\langle N(N-1)\rangle}{\bar{N}^2} = 1 + \frac{1}{\beta}.$$

This last result is identical with the result (4.17) (Chapter 4) as expected from the model. Nevertheless, the evolution to this stationary state does differ from that characterising the birth–death–immigration process, as can be seen in comparing Equation (6.20) with Equation (4.37).

Despite the death–multiple immigration and birth–death–immigration processes having different time dependences, the correlation functions have identical forms, albeit with different parameterisations, so that

$$G(\tau) \equiv \frac{\langle N(t)N(t+\tau)\rangle}{\langle N\rangle^2} = 1 + \left(\frac{1}{\beta} + \frac{1}{\bar{N}}\right)\exp(-\mu\tau). \tag{6.21}$$

Thus the characteristic timescale for the exponential decay of the number correlations is the inverse death rate and the fact that $G(\tau) \to 1$ as $\mu\tau \to \infty$ implies that the population sizes become uncorrelated at large separation times.

Note that taking the limit $\beta \to \infty$ in Equation (6.19) obtains the result

$$Q(s,t) = Q(s\theta, 0)\exp(-\bar{N}(1-\theta)s)$$

which is the generating function for the Poisson distribution for large times. This is consistent with the model because, from Equation (6.18), the limit $\beta \to \infty$ implies that $F(s) = -\mu \bar{N} s$, which corresponds to a single immigrant entering the population.

6.4 Sampling and Measurement Processes

In this section we examine the two measurement schemes introduced in Chapter 3 for the case when the population evolution is governed by the death–multiple immigration process having the negative binomial distribution as its stationary state. The procedure by which the evolution of the counting generating function is obtained according to how the measurements are enacted is now familiar, and so we can write down the partial differential equation for the generating function

$$\frac{\partial q}{\partial T} = -[s\bar{\mu}(z) - \eta z]\frac{\partial q}{\partial s} + F(s)q \tag{6.22}$$

where the form adopted by $\bar{\mu}(z)$ provides the distinction as to whether the monitoring is performed internally or externally:

$$\bar{\mu}(z) = \begin{cases} (\mu + \eta z) & \text{internal monitoring} \\ (\mu + \eta) & \text{external monitoring.} \end{cases}$$

The general solution of Equation (6.22) has the rather simple structure

$$q(s,z;T) = \bar{Q}_\infty(s'(s,z,T))\exp\{G(s,z) - G(s'(s,z,T),z)\} \tag{6.23}$$

where

$$s'(s,z,T) = \frac{\eta z}{\bar{\mu}(z)}(1 - \theta(T)) + s\theta(T), \quad \text{with} \quad \theta(T) = \exp(-\bar{\mu}(z)T) \tag{6.24}$$

and

$$G(s,z) = -\int_0^s dx \frac{F(x)}{(\bar{\mu}(z)x - \eta z)}. \tag{6.25}$$

$\bar{Q}_\infty$ is the stationary state solution of the population. Implicit in the solution (6.23) is that it is initiated from the stationary state that has been attained by the actual population, as represented by the function $\bar{Q}_\infty$, that is, the solution of Equation (6.4) with μ replaced with $\bar{\mu}(z)$, according to which monitoring scheme is used.

Clearly the explicit form of the solution (6.23) depends on the function F. For the case of negative binomial number fluctuations the form of the generating function is substantially different from the form resulting from the birth–death–immigration process (4.61), specifically

$$q(s,z;T) = \exp(\bar{N}\eta z[\kappa(z)-1]T)\left(1+\frac{\bar{N}}{\beta}\left(\frac{\eta z}{\bar{\mu}}(1-\bar{\theta})+s\bar{\theta}\right)\right)^{\bar{\beta}[\kappa(z)-1]}\left(1+\frac{\bar{N}}{\beta}s\right)^{-\bar{\beta}\kappa(z)} \tag{6.26}$$

where

$$\kappa(z) = \left(1+\frac{\bar{N}\eta z}{\bar{\mu}\beta}\right)^{-1} \quad \bar{\theta} = \exp(-\bar{\mu}T) \quad \text{and} \quad \bar{\beta} = \frac{a}{\bar{\mu}(1-\zeta)}.$$

The first thing to note about this solution is its dependence on the integration time T. When $T \to 0$, the solution reduces to Equation (6.17), which is the stationary state for the unmonitored population but with the death rate augmented by the additional factor η that characterises the monitoring process. As the integration time $T \to \infty$, the exponential factor in (6.26) dominates and the fluctuations are suppressed through being effectively averaged over many correlation times. The dependence on z in Equation (6.26) shows that the generating functions for the internal or external monitoring schemes differ substantially between themselves and from the birth–death–immigration process.

Despite these manifest differences, the form adopted by the first and second moments (and therefore the Fano factors) and first-order correlation functions of the monitored populations remains, remarkably, the same. Consequently, a programme to distinguish between the death–multiple immigration or birth–death–immigration processes based on these measures alone would fail. It is not difficult to demonstrate that third-order correlation coefficients, for example $\langle n^2(0)n(t)\rangle$, can be used to distinguish between the models, but a more effective way would be to use the times between the counted events during the course of external monitoring.

The distribution of the time between the events can be calculated from Equation (2.54) (see Chapter 2), which requires the generating function for the monitored population alone. This is obtained from Equation 6.26 with $\bar{\mu} = \mu + \eta$ by setting $s = 0$:

$$q_e(z;T) \equiv q_e(s=0,z;T) = \left(\frac{\kappa(z)}{\bar{\theta}^{-1}+\kappa(z)-1}\right)^{\bar{\beta}[1-\kappa(z)]}. \tag{6.27}$$

Differentiating twice with respect to T, and setting $z = 1$ gives

$$w_1(t) = \bar{\mu}\frac{\kappa^{\bar{\beta}(1-\kappa)+1}(1-\kappa)\bar{\theta}^{-1}\left(\bar{\beta}\bar{\theta}^{-1}+1\right)}{\left(\bar{\theta}^{-1}+\kappa-1\right)^{\bar{\beta}(1-\kappa)+2}}. \qquad (6.28)$$

Here $\kappa = \kappa(z = 1)$. Figure 6.4 shows the interevent distribution for a selection of parameter values; this is plotted as a cumulative probability, the probability that the interevent time is greater than t. This distribution is characterised by two time constants, $\bar{\mu}$ and $\bar{r}/(1+\bar{r}/\bar{\mu}\bar{\beta})$. The first characterises the bunching of events whilst the second relates to the mean event rate $\bar{r} = \eta\mu\bar{N}/\mu$ that can be derived from result (6.27).

Although the Fano factors for the birth–death–immigration and death–multiple immigration processes cannot be directly used as a mechanism to distinguish between them, the way that they feature in the interevent distributions can, especially if the tails of the density function are considered. If $\bar{\mu}T >> 1$ (which is certainly true for the tails of the interevent density), then F tends to an asymptotic limit given by

$$F = 1 + \frac{2\bar{N}\eta}{\beta\bar{\mu}} > 1.$$

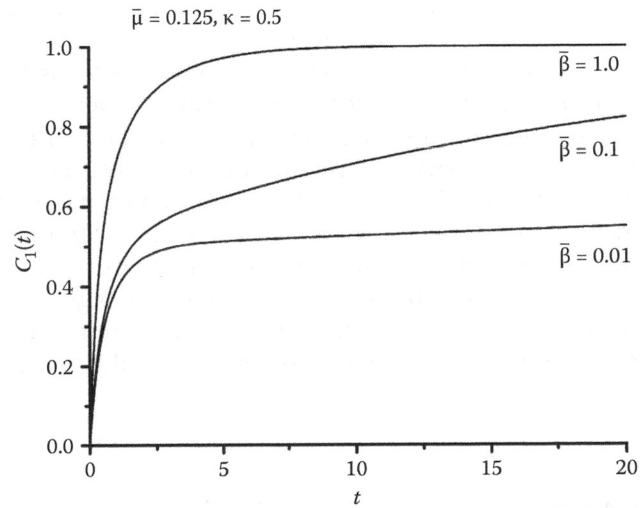

FIGURE 6.4
The cumulative distribution function for the interevent-time probabilities, which is the integral of Equation (6.28) from 0 to t. Note that, although the slope of the lower curve appears almost flat on this timescale, all of the probabilities tend to unity for sufficiently long times. This shows that it is possible to have two very different timescales in this process.

Now, because the generating functions for both processes have exponential tails, it follows that the interevent density function will also share the same asymptotic exponential dependence, that is,

$$w_1(\tau) = \frac{1}{\eta\bar{N}} \frac{\partial^2 q_e(z=1,T)}{\partial T^2}\bigg|_{T=\tau} \sim \exp(-\Theta\tau). \tag{6.29}$$

The inverse time constant Θ is different for the two processes, however

$$\Theta = \frac{\eta\bar{N}}{F-1}((2F-1)^{1/2}-1) \tag{6.30}$$

for the birth–death–immigration process and

$$\Theta = \frac{2\eta\bar{N}}{F+1} \tag{6.31}$$

for the death–multiple immigration process. Note that as $F \to 1$, the value of Θ for both models is consistent with that of the Poisson process, for which $\Theta = \eta\bar{N}$. For a fixed value of $F >> 1$, the exponent for the birth–death–immigration process is greater than that given for the death–multiple immigration process. Hence, fluctuations in the times between events occurring in the death–multiple immigration process show a greater spread than those for the birth–death–immigration process. This may have been anticipated since the events are driven by transitions involving a singleton alone for the birth–death–immigration process but by multiple transitions for the death–multiple immigration process, which can happen with a greater range of the rates. Figure 6.5 shows the dependence on F of the exponent of the interevent time density function in the asymptotic regime.

The exponent Θ has a wider currency, and is sometimes referred to as the 'persistence.' For example, the probability density for the time between flips of the elementary spins comprising a spin glass can have the asymptotic form given by Equation (6.29), and indeed these changes of state can be viewed as discrete events that are influenced by a more fundamental and complex dynamic whose precise nature is not necessarily known.

6.5 Time-Dependent Multiple Immigrations

We noted in Section 6.2 that the multiple immigration coefficients ν_r can be interpreted as probabilities and that the multiple immigration term in the rate equation for the generating function relating to the population of

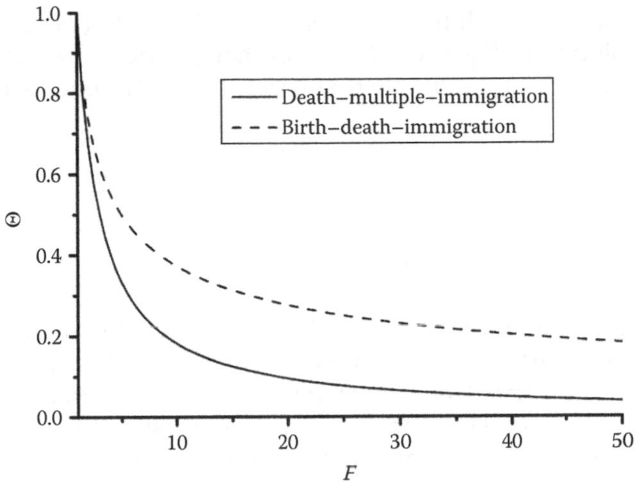

FIGURE 6.5
Results of Equations (6.30) and (6.31) for the time constant of the interevent density function of the monitored birth–death–immigration process and death–multiple immigration process, with $\eta\bar{N} = 1$.

interest could therefore be expressed in terms of the generating function $\hat{Q}$ for these probabilities according to Equation (6.7). This suggests a simple extension of the model to include variations in the immigrant probabilities so that $\hat{Q}$ becomes a function of time as well as the transform variable s and describes a process rather than a static distribution of immigrants. The more general problem of coupled populations will be considered in Chapter 8. However, it is worth considering here an example of the simpler case where the immigration process is just a driver for the population of interest.

Suppose that the immigrants are themselves governed by a thermal process:

$$\alpha\frac{\partial\hat{Q}}{\partial t} = -\mu s\frac{\partial\hat{Q}}{\partial s} + \lambda s(1-s)\frac{\partial\hat{Q}}{\partial s} - \lambda s\hat{Q}. \tag{6.32}$$

This is the same as Equation 4.11 (Chapter 4) with $\nu = \lambda$ except for a scaling of the time. If we assume that the population of interest is governed by the same thermal process but now with additional multiple immigrants, then the corresponding generating function is governed by the equation

$$\beta\frac{\partial Q}{\partial t} = -\mu s\frac{\partial Q}{\partial s} + \lambda s(1-s)\frac{\partial Q}{\partial s} - \lambda sQ + \varepsilon Q[\hat{Q}-1]. \tag{6.33}$$

By scaling the time differently in this equation we have ensured that the process exhibits fluctuations that are characterised by two distinct decay times. The stationary solution of Equation (6.32) is $\hat{Q} = (1+\bar{N}_2 s)^{-1}$ and the stationary solution to Equation (6.33) can then be found by substituting in this result and solving the simple first-order differential equation to give

$$Q_\infty = \frac{1}{1+\bar{N}_2 s}\exp\left[-\frac{\bar{N}_1 s}{1+\bar{N}_2 s}\right]$$
$$\bar{N} = \bar{N}_1 + \bar{N}_2. \tag{6.34}$$

Here, $\bar{N}_2 = \lambda/(\mu-\lambda)$ and $\bar{N}_1 = \varepsilon\bar{N}_2^2/\lambda$. This is identical to example (6.13) that, in the absence of births, required a more complicated structure for the multiple immigration term. The important feature here however is the time-dependent behaviour. This can be illustrated most simply by calculating the relaxation of the mean population by differentiating Equation (6.32) and Equation (6.33) with respect to s and setting $s = 0$:

$$-\alpha\frac{dN_2}{dt} = (\mu-\lambda)N_2 - \lambda$$
$$-\beta\frac{dN}{dt} = (\mu-\lambda)N - \lambda - \varepsilon N_2. \tag{6.35}$$

The solution of the first of these equations is

$$N_2(t) = \bar{N}_2 + (N(0)-\bar{N}_2)\exp[-(\mu-\lambda)t/\alpha]$$

whereas the solution of the second is

$$N(t) = \bar{N} + \left[N(0) - \bar{N} - \frac{\varepsilon\bar{N}_2(N_2(0)-\bar{N}_2)}{\lambda(1-\beta/\alpha)}\right]\exp\left[-\frac{\mu-\lambda}{\beta}t\right] +$$
$$+\frac{\varepsilon\bar{N}_2\left(N_2(0)-\bar{N}_2\right)}{\lambda(1-\beta/\alpha)}\exp\left[-\frac{\mu-\lambda}{\alpha}t\right].$$

It is immediately apparent that the relaxation of the mean from its value at the initial time is governed by decay on two distinct timescales and we might expect that the higher-order correlation properties and monitored statistics will also be characterised by two timescales.

Several other ways of exploiting multiple immigration in modelling the time evolution of classical populations can be envisaged. For example, the required behaviour could be obtained as a property of the *total* number of individuals in two populations that are linked by multiple immigration. Further examples will be considered in Chapter 8.

6.6 Summary

- In this chapter we have obtained a full solution of the death–multiple immigration process.
- We have shown how the multiple immigration coefficients can, in principle, be chosen so that the process generates an arbitrary equilibrium distribution.
- Limitations of the procedure have been highlighted by examining the particular case of a Laguerre equilibrium distribution. It was found that the parameter range over which the method could be used was restricted. A possible generalisation of the model to include births that might be more suitable in such cases was suggested and the general time-dependent solution for this case was given.
- We have shown how the multiple immigration coefficients can be chosen to reproduce the negative binomial probability distribution that was obtained as the equilibrium solution of the birth–death–immigration process of Chapter 4.
- We have shown that although the evolution of the populations in the birth–death–immigration and death–multiple immigration models is different, the second-order statistics are structurally identical so that measurements of these cannot be used to uniquely characterise or distinguish the models.
- The distribution of time between counts was found to be a more useful measure for the purpose of distinguishing the models. In particular the rate of its long separation–time exponential decay depends in a characteristic way on the Fano factor.
- We have demonstrated that the death–multiple immigrant process may be characterised by more than one time constant if the multiple-immigrant probabilities themselves constitute a time-dependent process.

Problems

6.1 Obtain expressions for the *time-dependent* mean and *equilibrium* variance of a death–multiple immigration process in terms of the function $F(s)$. Show that the autocorrelation function of the number fluctuations is given by

$$\langle N(0)N(t)\rangle = \bar{N}^2 + \bar{N}\exp(-\mu t) + \frac{1}{2\mu}\sum_{r=2}^{\infty} \nu_r r(r-1)\exp(-\mu t)$$

where

$$\bar{N} = \mu^{-1}\sum_{r=1}^{\infty} \nu_r r.$$

6.2 Find the driving function $F(s)$ in Equation (6.3) required to simulate the equilibrium population distribution of a birth–limited death process (Chapter 5, Section 5.4). Show that the model cannot be interpreted as a death–multiple immigration process.

6.3 Verify that selecting the multiple immigration rates as the terms in a geometrical series $\nu_r = a\zeta^r$ gives

$$F(s) = \frac{-a\zeta s}{(1-\zeta)^2(1+\zeta s/(1-\zeta))}.$$

Identify the constants a and ζ that make this expression consistent with the form that obtains the generating function for negative binomial number fluctuations.

6.4 Verify that the solution to the time-dependent multiple-immigration model with rates given by a geometrical progression $\nu_r = a\zeta^r$ is given by

$$Q(s,t) = Q(s\theta(t),0)\left(\frac{1+(s\bar{N}/\beta)\theta(t)}{1+(s\bar{N}/\beta)}\right)^{\beta}$$

where $\bar{N} = a\zeta/[\mu(1-\zeta)]$, and $\beta = a\zeta/[\mu(1-\zeta)^2]$, with $\theta(t) = \exp(-\mu t)$.

6.5 Show that the second factorial moment for the externally monitored death–multiple immigration process with negative binomial number fluctuations is given by

$$\langle n(n-1)\rangle = \langle n\rangle^2\left(1+\frac{1}{2\beta\gamma^2}(2\gamma+\exp(-\gamma)-1)\right)$$

where $\gamma = \bar{\mu}T/2$ and $\langle n\rangle = \eta\bar{N}T$. Deduce that when $\gamma >> 1$, the Fano factor

$$F(T) \approx 1+\frac{2\eta\bar{N}}{\beta\bar{\mu}}$$

and

$$q_e(z=1;T) \sim \left(\frac{2}{1+F}\right)^{\frac{F-1}{F+1}\bar{\beta}} \exp\left(-\frac{2\eta\bar{N}}{1+F}T\right).$$

Hence, deduce the asymptotic form of the interevent time probability density function.

Further Reading

E. Jakeman, K.I. Hopcraft and J.O. Matthews, 'Distinguishing population processes by external monitoring,' *Proceedings of the Royal Society, London*, **459**, 623–639 (2003).

J.O. Matthews, K.I. Hopcraft and E. Jakeman, 'Generation and monitoring of discrete stable random processes using multiple immigration population models,' *Journal of Physics A: Mathematical and General*, **36**, 11585–11603 (2003).

7

Stable Processes and Series of Events

7.1 Introduction

Although this book is concerned with discrete processes, the subject of the present chapter follows closely on the theory of stable *continuous* processes. The most commonly encountered model for continuous variables is based on the Gaussian probability density function. The reader is referred to the many texts on Gaussian variables and stochastic processes for details. Here we merely note that when independent Gaussian variables are added, then fluctuations in their sum are also characterised by a Gaussian probability density, a property known as statistical *stability*. In fact it has been known for more than 70 years that Gaussian is but one of a whole class of distributions governing the fluctuations of a variable with this property. In the case of continuous Gaussian noise, the property of statistical stability has greatly facilitated the mathematical description and analysis of a wide range of practical problems in all fields of science and engineering. However, other members of the class of stable distributions have recently begun to find applications on recognition of the importance of intermittency and other physical phenomena leading to outlying rare events. The relatively large probability of such events requires distributions with longer tails than Gaussian and other members of the stable class provide possible candidates for modelling data with such characteristics. These distributions have power-law tails $p(x) \sim 1/x^{1+\nu}$ with the index in the range $0 < \nu < 2$ so that the variance of the distributions is infinite. The special case $\nu = 2$ gives the more familiar Gaussian case $p(x) \propto \exp(-ax^2)$ for which all moments exist.

The property of long tails and the associated high probability of outlying rare events can also be important attributes of discrete distributions. Although mention was made of such distributions in Chapter 2, Section 2.2, none of the *processes* described in earlier chapters are characterised by long-tailed distributions. In this chapter, therefore, we shall develop a simple stochastic population model, based on multiple immigrations, for which the number probability distribution P_N decreases at large values of N in inverse proportion to the number raised to a power. This will enable us to calculate the evolutionary behaviour of a simple Markov population exhibiting extreme fluctuations in the number of individuals. Since the moments of the number distribution and the number correlation functions of the model do not exist, we shall investigate other measurable properties that will allow its fluctuation properties to be characterised.

7.2 Discrete Stability

We have already briefly encountered the concept of statistical stability in the context of discrete distributions in Chapter 2, Section 2.4, and it is useful to reproduce the discussion here. We begin by recalling that the moment generating function for the joint distribution of R discrete statistically independent random variables is defined by

$$
\begin{aligned}
Q(s_1, s_2, \ldots s_R) &= \\
&= \sum_{N_i=0}^{\infty} P(N_1, N_2, \ldots N_R)(1-s_1)^{N_1}(1-s_2)^{N_2} \ldots (1-s_R)^{N_R} \\
&= \prod_{n=1}^{R} Q^{(n)}(s_n).
\end{aligned}
\tag{7.1}
$$

The generating function $Q_\Sigma(s)$ for the *sum* of variables $N_1 + N_2 + \cdots N_R$ is obtained by setting $s_n = s$ in Equation (7.1) for all n. In the special case when the variables are also statistically identical so that $P^{(n)}(N_n) \equiv P(N_n)$ we then obtain

$$
Q_\Sigma(s) = Q^R(s). \tag{7.2}
$$

The distribution corresponding to Q is said to be *stable* if $Q_\Sigma(s) = Q(a_R s)$ where a_R is a constant, that is, if

$$
Q^R(s) = Q(a_R s). \tag{7.3}
$$

Thus the term *stable* refers to the property of distributions of sums of identical independent random variables that are essentially the same as those of the individual variables. The relation (7.3) is satisfied by generating functions that are exponentials of powers of s and these define positive definite distributions when

$$
Q(s) = \exp(-As^\nu) \qquad 0 < \nu \leq 1, \tag{7.4}
$$

where A is a positive constant. The Poisson distribution (2.2) evidently corresponds to the case $\nu = 1$ (see Equation 2.23) (see Chapter 2) and it is indeed well known that the sum of independent Poisson variables is also Poisson distributed with $\langle N\rangle = -dQ/ds\big|_{s=0} = A$. However, apart from this case all the members of the class (7.4) have infinite mean values, although the distributions themselves are positive and normalised. Moreover, they have power-law tails akin to the behaviour of the models (2.14) with $m = 0$ (see Chapter 2). To see this we first expand the exponent on the right-hand side of (7.4) using the binomial theorem (assuming that $0 < \nu < 1$):

$$As^{\nu} \equiv A[1-(1-s)]^{\nu} = A\sum_{n=1}^{\infty}\frac{\Gamma(n-\nu)}{\Gamma(-\nu)n!}(1-s)^n. \tag{7.5}$$

Now, the corresponding probability distribution is defined by the behaviour of the generating function (7.4) near $s = 1$, that is,

$$P_N = \frac{1}{N!}\left(-\frac{d}{ds}\right)^N Q(s)\big|_{s=1}. \tag{7.6}$$

Substituting Equation (7.4) into this expression and using Equation (7.5) to obtain the leading terms in powers of $(1 - s)$ leads, for large N, to

$$P_N \sim \frac{1}{N^{\nu+1}}. \tag{7.7}$$

7.3 A Discrete Stable Markov Process

In Chapter 6 we obtained the general solution of a Markovian death–immigration process in which immigrants entered a population not singly but in multiples and showed how some of the equilibrium distributions obtained in earlier chapters could be duplicated by a suitable choice for the number distribution for the immigrants. In the present context we need to find the choice of $F(s)$ in the equation for the generating function satisfying

$$\frac{\partial Q}{\partial t} = -\mu s\frac{\partial Q}{\partial s} + \left[\sum_{r=1}^{\infty}\nu_r\{(1-s)^r - 1\}\right]Q = -\mu s\frac{\partial Q}{\partial s} + F(s)Q \tag{7.8}$$

that will lead to the equilibrium distribution (7.4). The general equilibrium solution of Equation (7.8) is

$$Q(s,\infty) \equiv Q_{\infty}(s) = C\exp\left(-\int_0^s ds'\frac{F(s')}{\mu s'}\right). \tag{7.9}$$

When this is set equal to Equation (7.4) it is not difficult to show that $F(s)$ must have the form

$$F(s) = -A\mu\nu s^{\nu} = -as^{\nu}. \tag{7.10}$$

With this choice Equation (7.8) will characterise the evolution in time of a population having a stable equilibrium number distribution. However, it is important to recognise that this model represents just one particularly simple way in which a discrete stable Markov process may evolve. As we have seen in Chapter 6 other models giving the same first-order probability distribution may well predict a different time development.

The immigration rates corresponding to Equation (7.10) can be obtained by expanding the right-hand side of the equation in powers of $(1 - s)$ and then equating to powers in the sum in Equation (7.8) as indicated at the end of the last section. This gives

$$\nu_r = -a\frac{\Gamma(r-\nu)}{r!\Gamma(-\nu)}. \tag{7.11}$$

These coefficients are all positive only if $0 < \nu < 1$, which covers the whole parameter range for the power-law regime of the discrete stable distributions. Moreover, apart from a normalisation constant, the rates are independent of the death rate that is the only other parameter of the model. The special case $\nu = 1$ corresponds to no multiple immigrations, that is, $\nu_1 = 1$ and $\nu_r = 0$ for $r \geq 2$, which is the Poisson process. Figure 7.1 shows ν_r as a function of the order of the multiplets and reveals that the immigration rates themselves have power-law tails for large values of r, that is, $\nu_r \sim 1/r^{1+\nu}$; the power-law tails showing

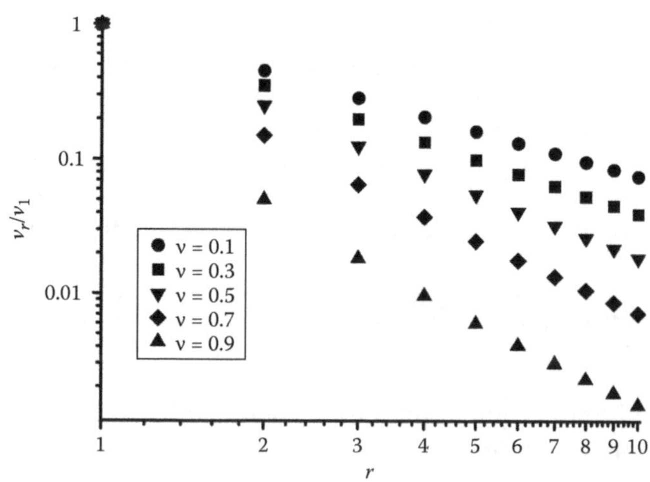

FIGURE 7.1
Multiple immigration rates for the discrete stable Markov process. The results for the different values of ν are normalised by the $r = 1$ rate.

as straight lines on this log–log plot. This demonstrates the dominance of the power-law behaviour and may be contrasted with the case when immigrant arrivals are governed by a geometrical distribution. We have seen in Chapter 6, Section 6.2, that the latter generates an entire class of distributions from one member of that class. The denominator of the integrand in Equation (7.9) is a consequence of deaths in the underlying process. Its presence ensures that for power-laws, and only for power-laws, the immigration rates are mirrored by the stationary distribution of the population itself.

The equation for the evolution of the generating function of a stable process generated by multiple immigrations is thus

$$\frac{\partial Q}{\partial t} = -\mu s \frac{\partial Q}{\partial s} - a s^{\nu} Q. \tag{7.12}$$

The time-dependent solution of this equation can be obtained following the Laplace transform technique described in Chapter 4, Section 4.4. The result is

$$Q(s;t) = \exp[-As^{\nu}(1-\theta^{\nu})]Q(s\theta;0). \tag{7.13}$$

The final factor on the right is just the generating function at $t = 0$ and will take the form $(1 - s\theta)^M$ if there are exactly M individuals present at the earlier time. The temporal coherence characterising the process is defined by the factor

$$\theta = exp(-\mu t). \tag{7.14}$$

After a long time $\theta \to 0$ and the result (7.13) approaches the desired equilibrium generating function (7.4). The evolution of the distribution corresponding to (7.13) is illustrated in Figure 7.2 for the case $\nu = 1/2$. The stationary solution for this case can be expressed in closed form in terms of a modified Bessel function of the second kind $K_\alpha(x)$ (see Glossary of Special Functions):

$$P_N = \frac{2}{N!\sqrt{\pi}}\left(\frac{1}{2}A\right)^{N+\frac{1}{2}} K_{N-\frac{1}{2}}(A). \tag{7.15}$$

Note that according to Figure 7.2 the tail of the distribution is established immediately, reflecting the fact that the moments of the distribution do not exist for any $t > 0$. This is because the rates ν_r permit a large number of immigrants to enter the population at a rate that is not exponentially bounded.

The joint generating function that describes the population having sizes N and N' following a delay time t can be deduced from the stationary solution together with Equation (7.13) conditioned on there being N individuals present initially:

$$Q(s,s';t) = \langle (1-s)^N (1-s')^{N'} \rangle = \sum_{N,N'=0}^{\infty} (1-s)^N (1-s')^{N'} P_N P(N'|N)$$

$$= \exp\left\{-\frac{a}{\nu\mu}[s'^{\nu}(1-\theta^{\nu}) + (s+(1-s)s'\theta)^{\nu}]\right\}. \tag{7.16}$$

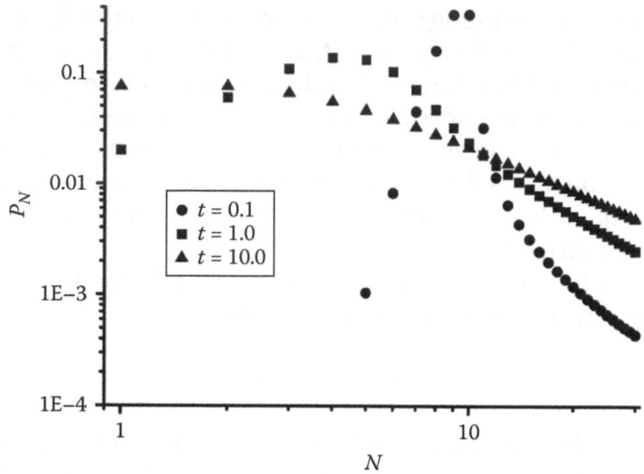

FIGURE 7.2
The time evolution of the probabilities for the death–multiple immigration process (from Equation 7.13). The values of the parameters are $a = 1.5$, $\nu = 0.5$ and $\mu = 1$. At $t = 0$ there are $M = 10$ individuals in the population.

All joint distributions and correlation properties can be obtained from this solution in principle because we are dealing with a first-order Markov process (see Equation 3.2, Chapter 3). However, because the joint probabilities also have power-law tails, the population number autocorrelation function is not defined. However, the characteristic evolution time embodied in θ defined by Equation (7.14) features in other measurable quantities that can be calculated from the model as we shall see later. A noteworthy property of Equation (7.16) is the absence of symmetry between s and s'. This implies that time ordering is a significant factor in any calculation relating to the model. For example, for some function $X(N(t))$ of the population number with finite mean and variance

$$\langle X(N(t))X^2(N(t'))\rangle \neq \langle X^2(N(t))X(N(t'))\rangle. \tag{7.17}$$

The rate Equation (7.8) based on deaths and multiple immigration is just one model that can be used to generate a stable population. Thus another one might be based on the *birth*–death–immigration process analysed in Chapter 4. When births are included in Equation (7.12) (see Chapter 4, Section 4.3) we obtain

$$\frac{\partial Q}{\partial t} = -s(\mu - \lambda(1-s))\frac{\partial Q}{\partial s} + F(s)Q, \tag{7.18}$$

where $F(s)$ is defined in Equation (7.8). It is not difficult to check that the general solution of this equation may be expressed in the form

$$Q(s;t) = Q_0(\Psi(s;t))\exp\left(-\int_0^t dt'\Psi(s;t')\right), \tag{7.19}$$

where

$$\frac{\partial \Psi}{\partial t} + s(\mu - \lambda(1-s))\frac{\partial \Psi}{\partial s} = 0$$
$$\Psi(s;0) = F(s). \tag{7.20}$$

As in the case of the birth–death–single immigration process a stationary solution exists only if the death rate exceeds the birth rate, $\mu > \lambda$, when

$$Q(s;\infty) = \exp\left(-\frac{1}{\mu-\lambda}\int_0^s ds' \frac{F(s')}{s'(1+bs')}\right), \tag{7.21}$$

with $b = \lambda/(\mu-\lambda)$. In order for this model to predict a stable equilibrium distribution, the form of $F(s)$ is different from Equation (7.10). In fact we must have

$$F(s) = A(\mu-\lambda)\nu(1+bs)s^{\nu} = a(1+bs)s^{\nu}. \tag{7.22}$$

The equivalent multiple immigration rates can be obtained by expanding this in powers of $(s-1)$ and equating the results to terms in the expansion (7.5):

$$\nu_m = \frac{a\Gamma(m-\nu)[m\lambda-(m-1-\nu)\mu]}{\Gamma(-\nu)m!(\mu-\lambda)(m-1-\nu)}. \tag{7.23}$$

Examples of these rates for a range of parameters are shown in Figure 7.3. Note that the rates depend on the birth and death rates in addition to the power law exponent ν, unlike the model based on deaths and multiple immigrations alone where they vary only with ν (Equation 7.11). In the case of the birth–death–multiple immigration model the generating function satisfies the equation

$$\frac{\partial Q}{\partial t} = -s(\mu-\lambda(1-s))\frac{\partial Q}{\partial s} - a\nu s^{\nu}(1+bs)Q. \tag{7.24}$$

The solution of this equation may be found using the method described in Chapter 4, which gives

$$Q(s;t) = \exp\left[\frac{-as^{\nu}}{(\mu-\lambda)\nu}\left(1-\frac{\theta_B^{\nu}}{[1+bs(1-\theta_B)]^{\nu}}\right)\right]Q\left(\frac{s\theta_B}{1+bs(1-\theta_B)};0\right), \tag{7.25}$$

where $\theta_B = \exp[-(\mu-\lambda)t]$. At long times an equilibrium solution is obtained of the form

$$Q(s;\infty) = \exp[-as^{\nu}/(\mu-\lambda)\nu], \tag{7.26}$$

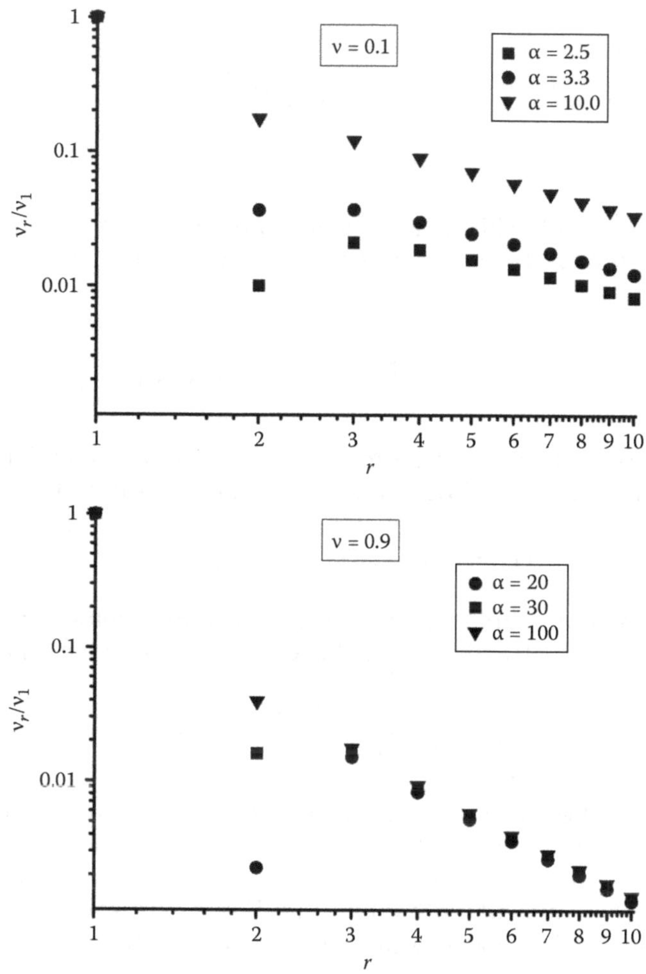

FIGURE 7.3
Normalised immigration rates for the birth–death–multiple immigration process (Equation 7.23). Here, α is the ratio of the death rate to the birth rate, that is, $\alpha = \mu/\lambda$. The two plots are for different values of the power-law coefficient ν.

which implies the desired asymptotic behaviour $P_N \sim N^{-(\nu+1)}$. Note that the immigration rates (7.23) are only positive provided that $0 < \nu < 1 - 2\lambda/\mu$, so that for a given value of μ and λ the full range of stable distributions cannot be generated by this model. However, for some range of parameters it will only be possible to assess the appropriateness of the death–multiple immigration and the birth–death–multiple immigration models through consideration of their time evolution, illustrated in Figure 7.4. The joint generating function governing this behaviour for the case of the

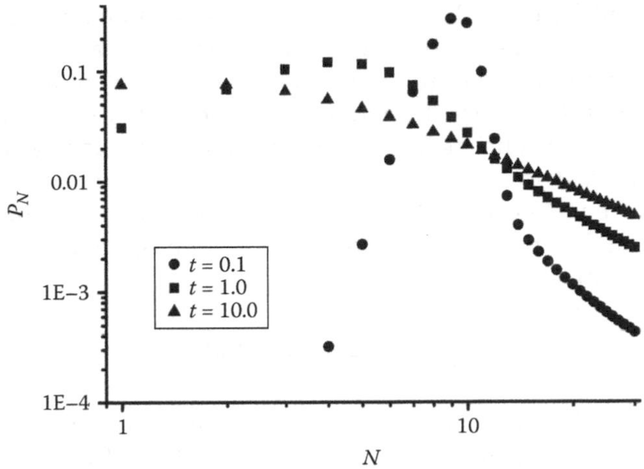

FIGURE 7.4
The time-dependent probabilities for the birth–death–multiple immigration process (from Equation 7.25), with $M = 10$ individuals at $t = 0$. The values of the parameters are chosen to give the same equilibrium distribution as the process in Figure 7.2, that is, $a = 1.5$, $\nu = 0.5$ and $\mu - \lambda = 1$. This still leaves a degree of freedom in the choice of λ, which has been set to 1/3 in this plot; this is the largest value allowed by the inequality $0 < \nu \le 1 - 2\lambda/\mu$. It turns out that the value of λ does not greatly affect the probabilities, which only differ slightly from those of Figure 7.2 in which $\lambda = 0$.

birth–death–multiple immigration process can be obtained by following the method of Chapter 4,

$$Q(s,s';t)$$

$$= \exp\left[-\frac{a}{\nu(\mu-\lambda)}\left(s'^{\nu}\left(1-\frac{\theta_B^{\nu}}{[1+bs'(1-\theta_B)]^{\nu}}\right)+\left(s+\frac{(1-s)s'\theta_B}{[1+bs'(1-\theta_B)]}\right)^{\nu}\right)\right]. \quad (7.27)$$

A result for the simpler death–multiple immigration process can be derived from this formula by setting the birth rate to zero and replacing θ_B with θ given by Equation (7.14).

7.4 Stable Series of Events

An interesting and potentially useful process that can be generated by the population model discussed in the last section is that described in earlier chapters in the context of *external* monitoring. As we have seen, this can be associated with the series of events generated by individuals *leaving* a population and here we shall assume that this population is governed by the above stable process. Following the approach of Chapter 4, Section 4.6, we imagine that removal of these individuals is equivalent to an additional death process in the internal population. As described in earlier chapters, an equation can be constructed that is satisfied by the generating function $q_e(s,z;T)$ for the distribution of the number of departures registered in a time interval T, when there were N present in the population at the beginning of the interval. For the simple death–multiple immigration model we find, defining the composite death rate $\bar{\mu} = \mu + \eta$,

$$\frac{\partial q_e}{\partial T} = (\eta z - \bar{\mu} s)\frac{\partial q_e}{\partial s} - as^{\nu} q_e. \tag{7.28}$$

The statistics of the number of events in a time interval T are obtained from the solution of this equation at $s = 0$. The full solution of Equation (7.28) can be found using the methods of Chapter 4, Section 4.6, and may be expressed in the form

$$q_e(s,z;T) =$$

$$Q(\Phi;\infty)\exp\left(-\frac{a}{(1+\nu)\eta z}\Phi^{\nu+1}{}_2F_1\left(1+\nu,1;2+\nu;{}^{\bar{\mu}\Phi}\!/_{\eta z}\right) - s^{\nu+1}{}_2F_1(1+\nu,1;2+\nu;\bar{\mu}s/\eta z)\right). \tag{7.29}$$

In this formula, ${}_2F_1(a,b;c;x)$ is a hypergeometric function (see Glossary of Special Functions) and $Q(\Phi;\infty)$ is the stationary solution (7.4) with μ replaced by $\bar{\mu}$ and Φ defined by

$$\Phi(s,z;T) = [\eta z + (\bar{\mu} s - \eta z)\exp(-\bar{\mu} T)]/\bar{\mu}. \tag{7.30}$$

A realisation of the process is shown in Figure 7.5. Note the variation in size and intermittent nature of the events. In fact the counting distribution has a power-law tail similar to the internal population distribution for all integration times as can be seen in Figure 7.6. This again expresses the stable property of the statistics: the events within an integration time that is much longer than the coherence time $\bar{\mu}^{-1}$ can be decomposed into many small groups, each of which is stably distributed and thereby so is their sum.

The generating function characterising the train of events formed by individuals leaving a population governed by a stable birth–death–multiple immigration process can also be obtained in closed form in terms of hypergeometric functions. The same methodology is applied that led to (7.29) but the results are somewhat more complicated and will not be reproduced here.

The generating function corresponding to the counting distribution alone is obtained from (7.29) by setting $s = 0$ and the counting distributions may be obtained in terms of the z-derivatives at $z = 1$ as described in Chapter 2, Section 2.3. The integration–time dependence of these quantities

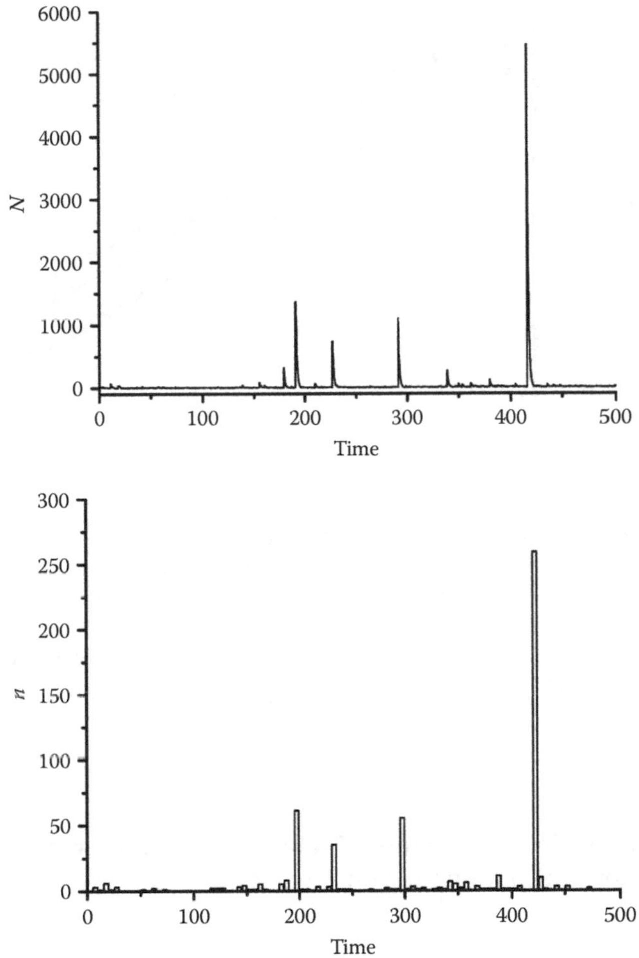

FIGURE 7.5
Death–multiple immigration process with external monitoring. Number in the population (upper plot) and number leaving the population during time intervals of duration $T = 5$ (lower plot). The values of the parameters are $a = 1.5$, $\nu = 0.7$, $\mu = 1$ and $\eta = 0.05$.

can in principle be used to obtain values for the parameters of the model. Unfortunately, as we have already noted, the *moments* of the number of events in fixed intervals are infinite and so, unlike all models described previously, the parameters characterising a stable series of events cannot be determined from these. However, other properties of the series of events can be calculated from result (7.29) including the distribution of times to the first event, $w_0(t)$, and the distribution of times between events, $w_1(t)$.

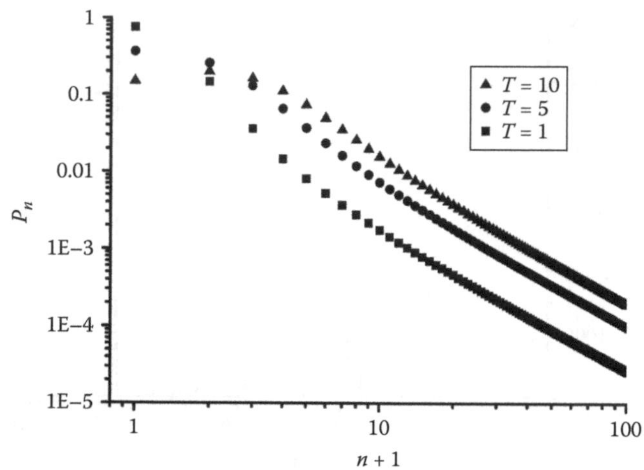

FIGURE 7.6
Probability distribution of the number of individuals counted leaving the population in the death–multiple immigration process with external monitoring. The parameters are the same as in Figure 7.5, with results shown for three different integration times T.

Setting $q(0,z;T) \equiv q(z;T)$, for simplicity, the time to the first event is given according to Equation (2.52) (Chapter 2) by

$$w_0(t) = -\frac{\partial q(1;t)}{\partial t}. \tag{7.31}$$

This quantity is plotted in Figure 7.7 and exhibits a power law dependence on t in the short integration time limit $\bar{\mu}t << 1$,

$$w_0(t) \sim a\left(\frac{\eta}{\bar{\mu}}\right)^{\nu}(\bar{\mu}t)^{\nu-1}. \tag{7.32}$$

For large times the events become uncorrelated and, as can be seen in Figure 7.7, the tail of the distribution is exponential as expected. The mean time to the first count is given by

$$\langle t_0 \rangle = \int_0^\infty t w_0(t)dt = \int_0^\infty p_0(t)dt. \tag{7.33}$$

Figure 7.8 shows the mean time to the first count as a monotonically increasing function of ν. It is readily confirmed that as $\nu \to 1$ it correctly approaches the asymptotic Poisson limit $(A\eta)^{-1} = \bar{\mu}/a\eta$ (see Section 7.2 and Chapter 2, Section 2.5). In this limit A is the mean number of individuals in the population and ηAT is the average number of events in the time interval of duration T.

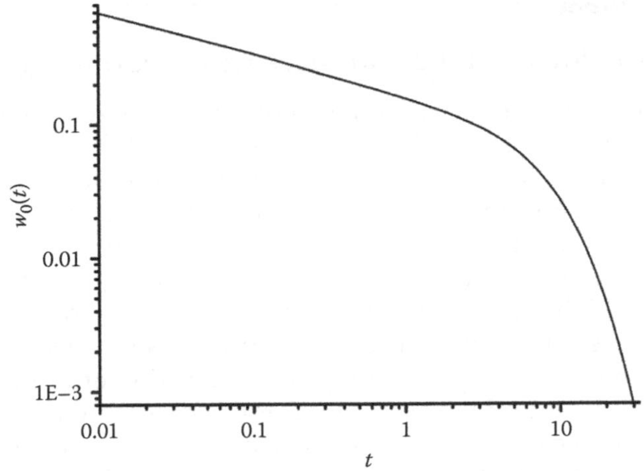

FIGURE 7.7
Probability distribution of the time to the first event for the monitored death–multiple immigration process. The parameters are the same as in Figures 7.5 and 7.6.

The derivation of the distribution $w_1(t)$ and mean time $\langle t_1 \rangle$ between events has to take into account the power-law behaviour of the generating function (7.29) with T at small integration times. As we shall see in the next section this implies that for the power-law population model, the apparent rate of occurrence of events *increases* with increasing resolution unlike the case of the simple death single-immigration process where it remains constant.

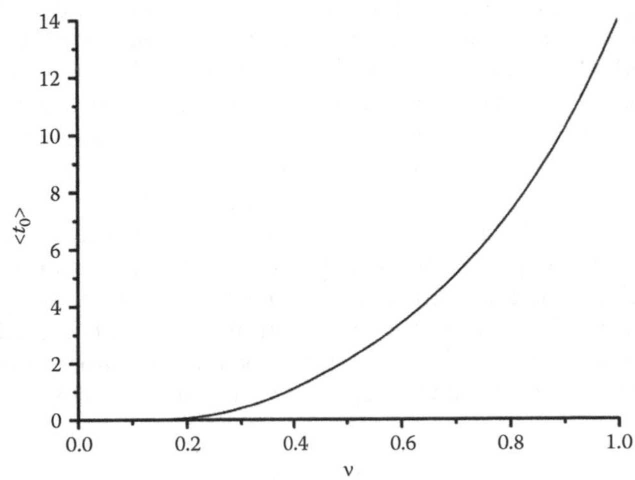

FIGURE 7.8
The mean time to the first count for the monitored death–multiple immigration process. The parameters are $a = 1.5$, $\mu = 1$ and $\eta = 0.05$.

7.5 Construction of Statistical Measures by Clipping

The characterisation of any real population process or series of events will be limited in practice by the dynamic range and the resolution of the measuring or detection system as well as by the finite time available for the measurement. These limitations are of particular concern in the present context for several reasons. In particular, the probability of finding a very large fluctuation in the number of individuals leaving the population is exceptionally high for power-law populations and the counter may saturate so that these events do not register correctly. Moreover, the chance of observing the most extreme events in an observation of finite duration is small. Hence, the tail of the event number distribution will be ill defined and statistical measures such as moments and correlation functions, though always finite, will change with measurement time. Thus the registered train of events will generally be different from that predicted by the ideal population model discussed in the previous sections, and although it may retain some characteristics of the original series, the loss of information during registration may lead to modified statistical properties and a reduction in the accuracy of parameter measurements. One method by which this problem can be controlled and quantified is to employ the technique of 'clipping' or 'limiting'.

Clipping is a technique more familiar in the context of analogue signal processing and was originally developed as a method for 'jamming' radar and communication systems. The term is generally applied to a process in which the original signal is replaced by a 'telegraph wave' that takes values ±1 according to whether the signal lies above or below a chosen level. In the case of Gaussian noise that is clipped at zero or *hard limited*, the spectrum is broadened such that the correlation function of the processed signal is related to that of the original by the arcsine formula or Van Vleck theorem, to be discussed further in Chapter 10, Section 10.2. The spectral spread achieved in this way is the required feature for jamming applications. A modified version of the technique was introduced in the late 1960s as a means of simplifying the post-detection processing of a series of photo-detection events in light scattering experiments. The quantity of interest in this case was the autocorrelation function of the event number and at the time this involved unattainably rapid multiplication of the number of events recorded in samples at different times. To overcome this problem the number of counts registered in each sample time was replaced by 1 or 0 as follows:

$$c_k(t) = \begin{cases} 0 & 0 \le n(t;T) \le k \\ 1 & n(t;T) > k. \end{cases} \tag{7.34}$$

It can be shown that the shape of the autocorrelation function of detected thermal light is not changed by clipping one autocorrelation channel in this way: a procedure that greatly simplifies the multiplication process at a small cost in statistical accuracy. Various refinements of the method, such as the use of a distribution of clipping levels and the technique of *scaling*, were developed to avoid distorting the correlation function in the case of other kinds of light. The interested reader is referred to the book on photon correlation spectroscopy and velocimetry listed under 'Further Reading' at the end of the chapter.

In view of the large excursions in count rate predicted for power-law population models, the procedure (7.34) might well be necessary in the present context to avoid the unknown non-linear effects of any counting method or device. It might also be useful as a simple model for such effects. Of course, application of the procedure will change both the statistics and correlation properties of the event stream. However, although the distribution of the new counting process will be different, it will have finite moments and correlation functions that will reflect the temporal evolution of the original population. At this point the question is whether the resulting statistical properties can be used effectively to determine the model parameters. In order to answer this question it is useful to consider the simplest but most extreme form of (7.34) when the series of events is hard limited:

$$c_k(t) = \begin{cases} 0 & n(t;T) = 0 \\ 1 & n(t;T) > 0. \end{cases} \tag{7.35}$$

Since the data now consists purely of 0 and 1 the generating function of the clipped distribution can be written as

$$q_c(z;T) = \sum_{n=0}^{1} (1-z)^n p_n(T) = p_0(T) + (1-z)(1-p_0(T)). \tag{7.36}$$

The mean clipped count rate is given by the first derivative with respect to z of this quantity evaluated at $z = 0$:

$$\langle c_0 \rangle = 1 - p_0(T) = 1 - q(1;T). \tag{7.37}$$

In the present case, the generating function on the right-hand side of this equation is given by Equation (7.29) and depends on the integration time. In fact, although the hard limiting process seems to be excessively severe, for sufficiently small times such that $\bar{\mu}T \ll 1$ the solution (7.29) predicts the asymptotic result

$$\langle c_0 \rangle \sim \frac{a}{\nu\bar{\mu}} (\eta T)^{\nu} + O(T^{2\nu}). \tag{7.38}$$

Thus at small sample times the mean number of clipped counts decreases more slowly with integration time than linear and the average time interval between these events will appear to increase with improved resolution as mentioned at the end of the previous section because $\langle t_1 \rangle = T/\langle c_0 \rangle \sim T^{1-\nu}$.

Measurement of the quantity (7.38) as a function of the integration time can be used to determine the power-law of the process and this shows that even the most severely limited integrated counting data preserve a remnant of the behaviour of the original series of events and hence of the internal population process. The use of more of the original data by clipping at a higher level, for example, will not necessarily simplify parameter determination but may allow more accurate measurements to be made in an experiment of the same duration. Note that clipping at zero is the simplest example of *saturation,* which occurs when the number of events up to a certain level is counted correctly but above the level the counter gives the same reading. For example, if the counter saturates at two events per counting interval then

$$q_{sat}^{(2)}(z;T) = p_0(T) + (1-z)p_1(T) + (1-z)^2(1 - p_0(T) - p_1(T)).$$

In this case the mean count rate is

$$\overline{c}_{sat}^{(2)} = 2[1 - p_0(T) - \tfrac{1}{2}p_1(T)].$$

It is also possible to calculate the autocorrelation function for the hard limited process (7.35) by using the joint distribution of counts in two non-overlapping integration periods of length T separated by a delay. Further discussion of this problem will be presented in Chapter 10. However, in the present context the procedure is quite lengthy and complicated and the interested reader is referred to the literature for results on this problem.

7.6 Summary

- This chapter dealt with discrete stable processes that have distributions with power-law tails, infinite moments and correlation functions.
- We have briefly reviewed the concept of statistical stability in the context of continuous variables and introduced the discrete stable distributions with reference to the discussion in Chapter 2.
- The multiple immigration rates required together with deaths to generate a discrete stable Markov process were then derived and found to be stably distributed themselves.
- A solution for the generating function of the joint distribution of the number in the population at two different times was found, and

some results were also derived for a model including births as well as deaths and multiple immigration.

- An equation for the generating function of the joint distribution of individuals in the population and those leaving it was solved and it was shown that the leaving event interval statistics reflect the power-law behaviour of the internal population.
- The concept of clipping the raw event data was introduced in order to generate a new discrete process that could be characterised by finite moments and correlation functions.
- The new process was shown to retain important features of the population statistics and could provide a useful data processing technique.

Problems

7.1 The number n of events in a time T is the sum of N identical statistically independent Poisson distributed variables, each derived from the same constant rate $\bar{r}$. Show that n is also Poisson distributed and find its mean value. Write an expression for the probability of finding no events in a time interval of duration t and hence find the probability distribution of the time to the first event $w_0(t)$.

7.2 Express the generating function for a stable distribution of index ½ as a modified Bessel function of the second kind. Using the differentiation formulae for Bessel functions, derive the corresponding discrete probabilities.

7.3 Derive the time-dependent generating function for the stable distribution governed by the death–multiple immigration process (Equation 7.8). Find an expression for the probability of finding no individuals in the population after a time t given that there were none present initially. Show that the time taken to achieve half of the decrease of P_0 towards its equilibrium value is $-\mu^{-1}\ln\{(\mu/a)\ln\frac{1}{2}[1+\exp(a/\mu)]\}$. Prove that this time is independent of μ if $a >> \mu$.

7.4 Show that the second factorial moment of a data stream consisting of only zeros and ones vanishes, and that the Fano factor is less than unity. Find expressions for the Fano factor of the signals generated by hard limiting (1) a random series of events, (2) events generated by individuals leaving a thermal population and (3) individuals leaving the death–immigration power-law population. Show that in the large integration time limit the Fano factor vanishes in the case of examples (1) and (3) but approaches a finite limit in the case of example (2).

7.5 Show how the mean and second moment of data measured by a counter that saturates when two or more events are counted in each sample can be used to deduce the probability of finding one or no counts in samples of the original data. Evaluate the sample time dependence of the generating function for the original distribution in the limit of small sample time and obtain expressions for the mean and variance of the measured data in this limit for the case of a stable death–multiple immigration process. Prove for this model that the Fano factor of the measured data in the short sample time limit can be used to determine the power-law index and is given by $F = (4 - 3\nu)/(2 - \nu)$.

Further Reading

E. Jakeman, *Photon Correlation in Photon Correlation and Light Beating Spectroscopy*, New York, Plenum Press, 1974.

P. Lévy, *Théory de l'Addition des Variables Aléatories*, Paris, Gauthier-Villars, 1937.

S.B. Lowen and M.C. Teich, *Fractal-Based Point Processes*, New York, Wiley, 2005.

J.O. Matthews, K.I. Hopcraft and E. Jakeman, 'Generation and monitoring of discrete stable random processes using multiple immigration population models,' *Journal of Physics A: Mathematical and General*, **36**, 11585–11603 (2003).

G. Samorodnitsky, and M.S. Taqqu, *Stable Non-Gaussian Random Processes: Stochastic Models with Infinite Variance*, New York, Chapman & Hall, 1994.

T.J. Shepherd and E.J. Jakeman, 'Statistical analysis of an incoherently coupled, steady-state optical amplifier,' *Journal of the Optical Society of America B*, **4**, 1860–1869 (1987).

J.H. Van Vleck and D. Middleton, 'The spectrum of clipped noise,' *Proceedings of the IEEE*, **54**, 2–19 (1966).

8

Coupled Processes

8.1 Introduction

In Chapter 6 we developed a simple Markov process based on multiple immigrations and in Chapter 7 we showed how this model could be exploited to describe the time evolution of a stable variable and generate a stable series of events—characteristics often ascribed to complex systems. The main part of this chapter develops the multiple immigrations model further, based on the considerations of Chapter 6, Section 6.5. There we noted that the multiple immigration coefficients ν_r could themselves be interpreted as probabilities so that the multiple immigrations term in the rate equation for the generating function $Q(s,t)$ relating to the population of interest could therefore be expressed in terms of the generating function $\hat{Q}(s)$ for the multiple immigration probabilities. One consequence of this was that temporal variations in the immigration coefficients could be included in the model by ascribing a time evolution to $\hat{Q}$ through a second process. Thus the number of immigrants arriving in the population of interest is effectively governed by the evolution of an *ensemble* of secondary populations. It is of course possible for the secondary populations to be influenced in the same way by an ensemble of primary ones. Here we shall investigate the consequence of this type of two-way coupling through study of simple death–multiple immigration models. We shall show that models of this kind are exactly solvable under certain conditions and we shall explore their potential for generating stable populations and oscillatory behaviour.

In the last section of the chapter we shall investigate a different kind of coupling between two populations where individuals leave each population through a death process and appear as immigrants in the other population, independent of the number present in that population at the time. This *exchange interaction* model is in general a much more difficult problem to solve, but again we shall find that in the case of two simple death–immigration processes it can be solved exactly and we shall investigate how the introduction of coupling changes the original populations. This kind of process can be used to model simple chemical reactions where one reactant is changing into the other and vice versa, and there are a number of biological systems that have been studied where the dynamics of coupled systems is relevant. An important aspect of such problems that we shall also investigate here is the effect of the measurement process.

8.2 Coupled Death–Multiple Immigrations Processes

With the aforementioned considerations in mind, according to Chapter 6 we can write the equations for the generating functions for two death processes coupled by multiple immigrations in the form

$$\begin{aligned}\frac{\partial Q^{(1)}}{\partial t} &= -\mu_1 s \frac{\partial Q^{(1)}}{\partial s} + \varepsilon_1 Q^{(1)}[Q^{(2)} - 1] \\ \frac{\partial Q^{(2)}}{\partial t} &= -\mu_2 s \frac{\partial Q^{(2)}}{\partial s} + \varepsilon_2 Q^{(2)}[Q^{(1)} - 1].\end{aligned} \quad (8.1)$$

It is important to emphasise that in this model members of the two populations are not exchanged nor are necessarily of the same type, but the number in one population merely affects the likelihood of a particular number of immigrants arriving in the other as indicated in Figure 8.1. Unfortunately, the model described by (8.1) generally leads to runaway populations or

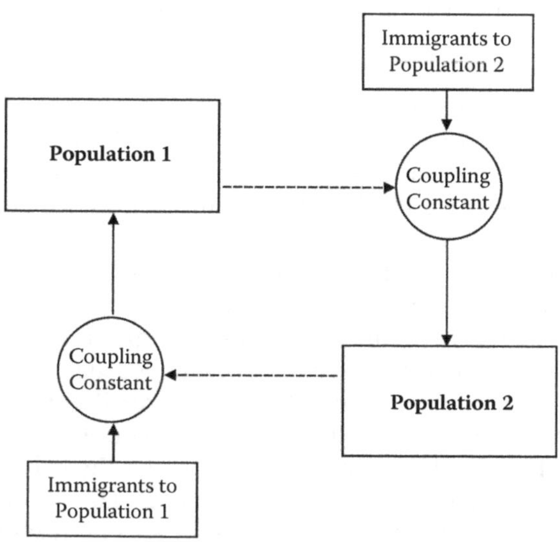

FIGURE 8.1
Coupled death–multiple immigration process.

populations that become extinct. To see this, simply differentiate the two equations with respect to s and set $s = 0$ to obtain time-dependent equations for the means:

$$\frac{d\bar{N}}{dt} = -\mu_1\bar{N} + \varepsilon_1\bar{M}$$

$$\frac{d\bar{M}}{dt} = -\mu_2\bar{M} + \varepsilon_2\bar{N}.$$

It is not difficult to demonstrate that there is no non-zero, time-independent solution of these equations except in the special case $\mu_1\mu_2 = \varepsilon_1\varepsilon_2$. In order to retain some generality in the model, therefore, it is useful to include a separate single immigrant term in each of the equations of (8.1):

$$\begin{aligned}\frac{\partial Q^{(1)}}{\partial t} &= -\mu_1 s\frac{\partial Q^{(1)}}{\partial s} + \varepsilon_1 Q^{(1)}[Q^{(2)} - 1] - \nu_1 s Q^{(1)} \\ \frac{\partial Q^{(2)}}{\partial t} &= -\mu_2 s\frac{\partial Q^{(2)}}{\partial s} + \varepsilon_2 Q^{(2)}[Q^{(1)} - 1] - \nu_2 s Q^{(2)}.\end{aligned} \tag{8.2}$$

This ensures that finite equilibrium distributions exist. In the absence of coupling the equations describe independent death–immigration processes characterised by Poisson equilibrium solutions and the associated events generated by individuals leaving each population (external monitoring) are uncorrelated as discussed in Chapter 3, Section 3.5.

In the presence of coupling the solution of the equations is more complicated and we shall specialise to the case $\mu_1 = \mu_2 = \mu$, $\varepsilon_1 = \varepsilon_2 = \varepsilon$, $\nu_1 = \nu_2 = \nu$. This will enable us to obtain full solutions for the generating functions, although it is possible to solve the more general case for the lower moments by finding the derivatives of the equations with respect to s at $s = 0$.

When the parameters of the model are identical as indicated earlier, the non-linear term can be removed by subtraction to give

$$\begin{aligned}\frac{\partial D}{\partial t} &= -\mu s\frac{\partial D}{\partial s} - (\varepsilon + \nu s)D \\ D &= Q^{(1)} - Q^{(2)}.\end{aligned} \tag{8.3}$$

Using the methods described in earlier chapters, a solution of the first equation can be obtained in the form

$$D(s,t) = D(s\theta,0)\exp\left[-\frac{\nu s}{\mu}(1-\theta) - \varepsilon t\right] = D(s\theta,0)f(s,t), \tag{8.4}$$

where $\theta(t) = \exp(-\mu t)$. For example, if there were N_0 individuals present in population 1 and M_0 in population 2 initially then

$$D(s\theta,0) = (1-s\theta)^{N_0} - (1-s\theta)^{M_0}. \tag{8.5}$$

The second equation of (8.3) may be used to eliminate $Q^{(2)}$ from (8.2). The substitution $Q^{(1)} = D/[1-\exp(X)]$ then reduces this equation to the standard form

$$\frac{\partial X}{\partial t} + \mu s \frac{\partial X}{\partial s} = \varepsilon D. \tag{8.6}$$

The particular integral of this equation can be obtained in terms of the initial value of X by Laplace transformation with respect to the time variable. It is then necessary to add the complementary function so as to satisfy the boundary condition at the initial time. The full solution is found to be

$$Q^{(1)} = \frac{Q_0^{(1)} D_0 f}{Q_0^{(1)} - Q_0^{(2)} \exp[D_0(1-f+R)]}. \tag{8.7}$$

Here $D_0 = D(s\theta,0)$, $f = f(s,t)$ as shown in Equation (8.4) and R is defined in terms of the incomplete gamma function, $\Gamma(a,x) = \Gamma(a) - \gamma(a,x)$, by the formula

$$R(s,t) = \left(\frac{\nu s}{\mu}\right)^{\varepsilon/\mu} \exp\left(\frac{\nu s}{\mu}\theta - \varepsilon t\right)\left[\Gamma\left(1-\frac{\varepsilon}{\mu}, \frac{\nu s}{\mu}\right) - \Gamma\left(1-\frac{\varepsilon}{\mu}, \frac{\nu s}{\mu}\theta\right)\right]. \tag{8.8}$$

The generating function for the second process $Q^{(2)}(s,t)$ can be obtained from the second equation of (8.3).

8.3 Properties of Populations Coupled by Multiple Immigrations

We first note that in the absence of the single immigrant terms, $\nu = 0$, the solution (8.7) reduces to

$$Q^{(1)}(s,t) = \frac{Q_0^{(1)} D_0 \exp(-\varepsilon t)}{Q_0^{(1)} - Q_0^{(2)} \exp\{D_0[1-\exp(-\varepsilon t)]\}}.$$

The corresponding result for population 2 can be obtained from the second equation of (8.3). Assuming that there were N_0 individuals present in population 1 and M_0 in population 2 initially, then according to this result the predicted means as a function of time are given by

$$\bar{N}(t) = \tfrac{1}{2}(N_0 + M_0)\exp[(\varepsilon - \mu)t] + \tfrac{1}{2}(M_0 - N_0)\exp[-(\varepsilon + \mu)t],$$
$$\bar{M}(t) = \tfrac{1}{2}(N_0 + M_0)\exp[(\varepsilon - \mu)t] + \tfrac{1}{2}(N_0 - M_0)\exp[-(\varepsilon + \mu)t]. \tag{8.9}$$

As time increases, both populations become extinct if $\varepsilon < \mu$ and increase without limit if $\varepsilon > \mu$. In the special case $\varepsilon = \mu$ the distribution of individuals in each population becomes the same at long times and the total number of individuals present initially is preserved with

$$Q_\infty^{(1,2)} = \frac{1}{1 + \tfrac{1}{2}(N_0 + M_0)s}.$$

Thus, in this case, the coupled multiple immigration process remembers the initial condition and a comparison with Equation (2.27) (see Chapter 2) shows that each population is governed by a negative binomial distribution of unit index or *geometric* distribution. Note, however, that the two populations are not necessarily made up from the same kind of individual. When $\nu \neq 0$ with $\varepsilon = \mu$ it can be shown that in the long time limit the solution (8.7) exhibits the following asymptotic behaviour:

$$\lim_{t\to\infty} Q^{(1)}(s,t) \to \frac{1}{1 + s\exp(\nu s/\mu)[\nu t + \tfrac{1}{2}(N_0 + M_0)]}. \tag{8.10}$$

In this case the means grow linearly with time at long times so that the solution is not strictly stationary. However, it turns out that the *normalised* factorial moments are independent of time in this limit, taking the values $r!$ expected for a geometric distribution.

Insight into the more general case when $\mu \neq \varepsilon$ is obtained by calculating the low moments of each process directly from the rate Equations (8.2). Denoting the mean and r^{th} factorial moment by $\bar{N}$ and $N^{[r]}$ as usual we find for the means and second factorial moments

$$\frac{d\bar{N}}{dt} = -\mu\bar{N} + \varepsilon\bar{M} + \nu, \quad \frac{dN^{[2]}}{dt} = -2\mu N^{[2]} + \varepsilon M^{[2]} + 2\bar{N}(\varepsilon\bar{M} + \nu),$$
$$\frac{d\bar{M}}{dt} = -\mu\bar{M} + \varepsilon\bar{N} + \nu, \quad \frac{dM^{[2]}}{dt} = -2\mu M^{[2]} + \varepsilon N^{[2]} + 2\bar{M}(\varepsilon\bar{N} + \nu). \tag{8.11}$$

Observe that the non-linear coupling in the model is not manifest in the equations for the mean values but first appears in the equations that determine properties of the fluctuations. Thus the dynamics of the mean of the population is not a good guide to the whole range of its behaviour. This is an important observation, since 'mean field' approximations are often used to expedite the description of processes that are intrinsically discrete. We also note that, as in the case of all models investigated previously, the properties of the higher moments of the distribution can always be expressed in terms of those of the lower ones so that no closure requirement has to be applied to obtain a solution for this system. When there are N_0 individuals present in population 1 and M_0 in population 2 initially, the evolving means of the two populations are given by

$$\begin{aligned}\bar{N}(t) &= \bar{N}_\infty(1-\exp[(\varepsilon-\mu)t]) + \exp(-\mu t)[N_0 \cosh \varepsilon t + M_0 \sinh \varepsilon t],\\ \bar{M}(t) &= \bar{N}_\infty(1-\exp[(\varepsilon-\mu)t]) + \exp(-\mu t)[M_0 \cosh \varepsilon t + N_0 \sinh \varepsilon t],\end{aligned} \tag{8.12}$$

where

$$\bar{N}_\infty = \nu/(\mu-\varepsilon). \tag{8.13}$$

In the absence of stabilisation ($\nu = 0$) these results reduce to Equation (8.9), whereas if $\nu \neq 0$ then $\bar{N}_\infty$ is the asymptotic mean of both populations at long times provided that $\mu > \varepsilon$. In this case the normalised second factorial moment is given by

$$N^{[2]} = \frac{2\mu}{2\mu-\varepsilon}. \tag{8.14}$$

This result approaches that expected for a Poisson distribution when there is no multiple immigration and approaches that of a geometric distribution as the two parameters become equal. Indeed, as mentioned earlier, in this special case all the higher factorial moments approach those expected for a geometric distribution though with a mean value νt that increases linearly with time. Figure 8.2 shows the evolution of the means in Figures 8.2a,c and the normalised second factorial moments in Figures 8.2b,d with time predicted by Equations (8.11) in the case $\mu > \varepsilon$. The behaviour for two values of the ratio ε/μ assuming that $N_0 = 5$ and $M_0 = 10$ is shown. The behaviour in both cases is broadly similar. The means rapidly converge, whilst the second moments rise to a peak before subsiding to the asymptote predicted by Equation (8.14). This effect is larger for the population starting from the lower initial number. The peaks are more pronounced and the approach to equilibrium is slower when ε/μ takes a value closer to unity.

The time-delayed auto- and cross-correlation functions for the two populations can be calculated directly from results (8.12) for the means by averaging over the initial states as described earlier (see Chapters 3 and 4, for example). The normalised correlation coefficients can be expressed in the form

$$\begin{aligned}G_{11}(\tau) = G_{22}(\tau) &= 1-\exp(-\mu\tau)\sinh(\varepsilon\tau) + \frac{\operatorname{var} N}{\bar{N}_\infty^2}\exp(-\mu\tau)\cosh(\varepsilon\tau)\\ G_{12}(\tau) = G_{21}(\tau) &= 1-\exp(-\mu\tau)\sinh(\varepsilon\tau) + \frac{\operatorname{var} N}{\bar{N}_\infty^2}\exp(-\mu\tau)\sinh(\varepsilon\tau)\end{aligned} \tag{8.15}$$

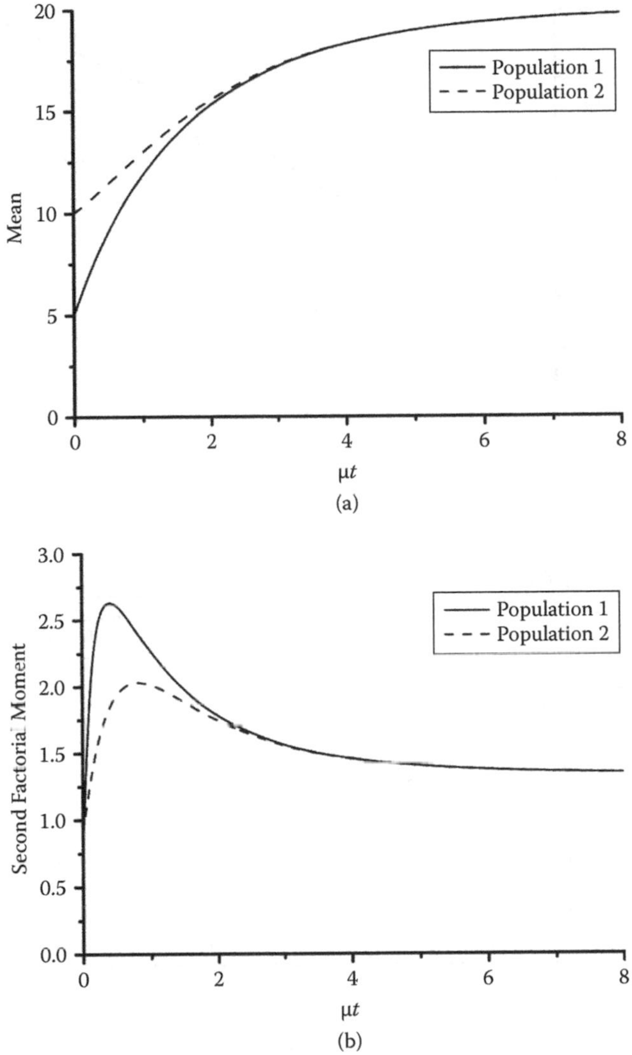

FIGURE 8.2
Evolution of the mean and normalised second factorial moment of the coupled population process. Here, $\nu = 10\mu$, $N_0 = 5$ and $M_0 = 10$. For (a) and (b) the coupling constant is $\varepsilon = 0.5\mu$ and for (c) and (d) it is $\varepsilon = 0.9\mu$. *(Continued)*

where the subscripts refer to the two populations and the relative variance of N is determined by Equations (8.13) and (8.14). Like the evolution of the means of the populations, these are characterised by two timescales $(\mu - \varepsilon)^{-1}$ and $(\mu + \varepsilon)^{-1}$. The presence of slow fluctuations when ε is almost equal to μ is revealed in Figure 8.3.

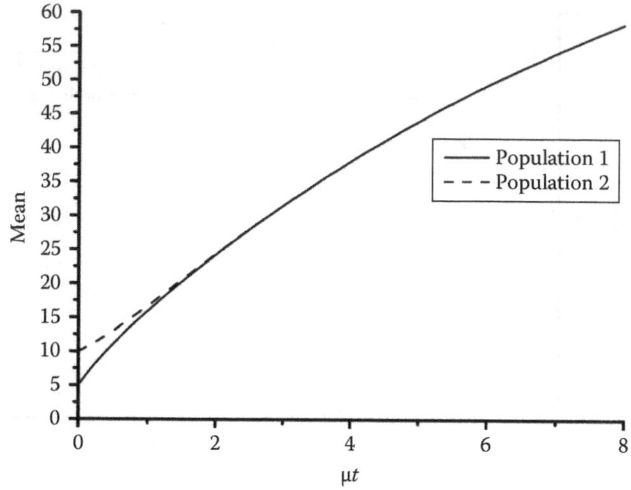

(c)

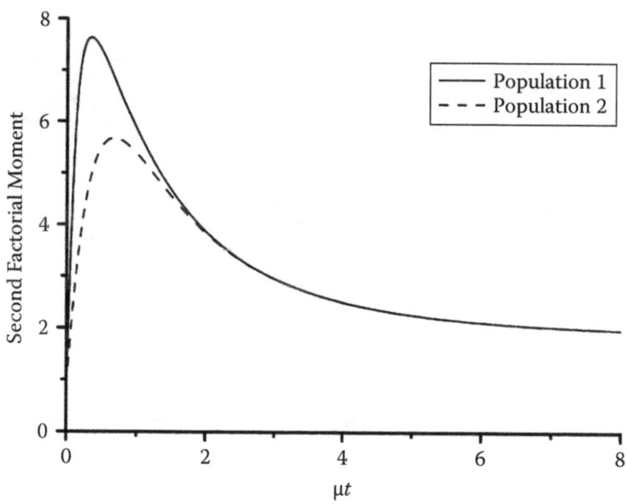

(d)

FIGURE 8.2 ***(Continued)***
Evolution of the mean and normalised second factorial moment of the coupled population process. Here, $\nu = 10\mu$, $N_0 = 5$ and $M_0 = 10$. For (a) and (b) the coupling constant is $\varepsilon = 0.5\mu$ and for (c) and (d) it is $\varepsilon = 0.9\mu$.

Solution (8.7) is valid for all initial states of the two populations. Suppose then that the initial state of population 2 is stable, $Q^{(2)}(s,t=0)=\exp(-as^{\alpha})$ with $0 < \alpha < 1$. This means that in Equation (8.7), $Q_0^{(2)} = \exp[-as^{\alpha}\exp(-\alpha\mu t)]$. Examination of the large time limit of the solution reveals that a stationary state still exists but now over the restricted range $\alpha\mu \geq \varepsilon$. When $\alpha\mu = \varepsilon$ the

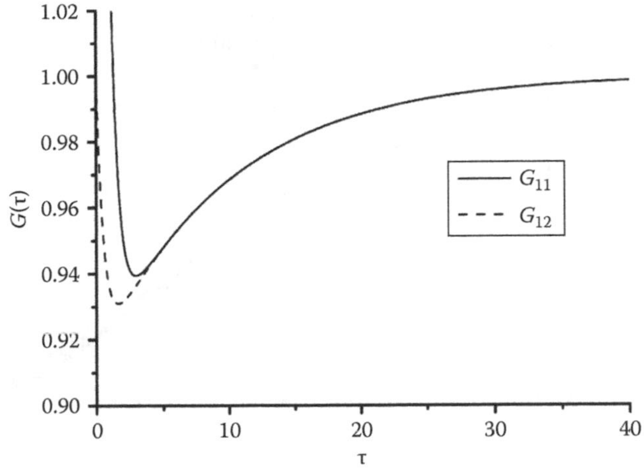

FIGURE 8.3
Auto- and cross-correlation functions for the coupled populations. Parameters are $\nu = 10$, $\mu = 1$ and $\varepsilon = 0.9$.

result is most compactly expressed in terms of the incomplete gamma function $\gamma(a,x) = \Gamma(a) - \Gamma(a,x)$:

$$Q_{\infty}^{(1)}(s) = \frac{1}{1+\exp(\nu s/\mu)\left[\frac{1}{2}as^{\alpha} + (\nu s/\mu)^{\alpha}\gamma(1-\alpha, \nu s/\mu)\right]}. \tag{8.16}$$

This result is truly stationary, unlike the case when there are a fixed number of individuals present initially (8.10). However, when $\nu = 0$ a stationary solution exists only when $\alpha\mu = \varepsilon$ and takes the form

$$Q_{\infty}^{(1)}(s) = [1 + as^{\alpha}/2]^{-1}. \tag{8.17}$$

The distributions generated by Equations (8.16) and (8.17) decrease with tails having the same power law as the initial distribution $\sim N^{-(\alpha+1)}$. As a consequence their integer moments are infinite but, unlike the initial distribution, they are *not* stable. Figure 8.4 shows plots of the distribution corresponding to Equations (8.16) and (8.17) compared with the initial stable one. In fact the coupled population model converts any initial population governed by a distribution with a power-law tail of the form $N^{-(\alpha+1)}$ with $0 < \alpha < 1$ into one with the distribution (8.17) provided that $\alpha\mu = \varepsilon$. Although (8.17) is not stable, it is a stationary invariant of the convolution Equation (6.1). In other words, if the static multiple immigration coefficients $\{\nu_r\}$ are probabilities corresponding to (8.17) then the stationary solution of (6.1) is also of this form. Moreover, if the time-dependent terms in Equation (8.1) are discarded, then a 'stationary' solution of the form (8.17) with $\alpha\mu = \varepsilon$ is predicted if it is

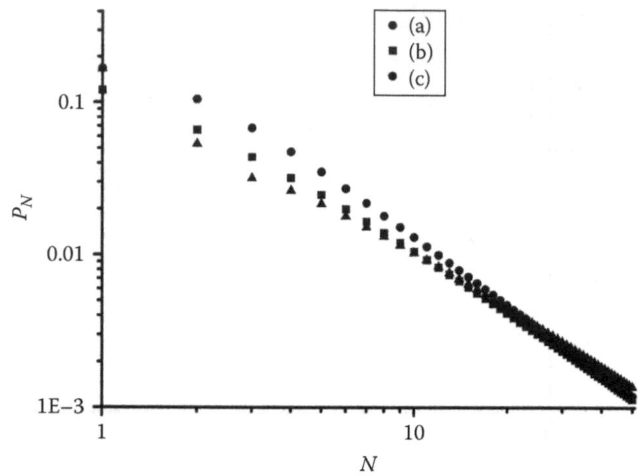

FIGURE 8.4
Probability distributions of the coupled populations. (a) The initial population, with $a = 1.5$ and $\alpha = 0.5$. (b) Distribution derived from Equation (8.17). (c) Distribution derived from Equation (8.16), with $\nu = 0.5\mu$.

assumed that $Q^{(1)} = Q^{(2)}$. However, this special class of solutions cannot be accessed from an arbitrary initial condition.

8.4 Cyclically Coupled Populations

The model described by Equation (8.2) can be extended to include more populations and various cross coupling schemes can be envisaged. As an example we consider in this section a simple configuration obtained by linking the populations in a chain and then connecting the ends to form a cyclic arrangement. This model is characterised by the rate equation

$$\frac{\partial Q^{(i)}}{\partial t} = -\mu_i s \frac{\partial Q^{(i)}}{\partial s} + \varepsilon_i Q^{(i)}[Q^{(i+1)} - 1] - \nu_i s Q^{(i)}. \tag{8.18}$$

Here, $i = 1, 2, 3, \ldots, n$ and $Q^{(n+1)} = Q^{(1)}$. In the case $n = 3$, for example, a solution for the average number in each population as a function of time can easily be derived. The simplest result is obtained when the death rates, single immigration rates, and coupling constants are the same for each population:

$$\bar{N}_i(t) = \bar{N} + \tfrac{1}{3}(N_1 + N_2 + N_3 - 3\bar{N})\exp[(\varepsilon - \mu)t] +$$
$$+ \tfrac{1}{3}\left[(2N_1 - N_2 - N_3)\cos(\varepsilon t\sqrt{3}/2) + (N_2 - N_3)\sin(\varepsilon t\sqrt{3}/2)\right]\exp[-(\mu + \varepsilon/2)t]. \tag{8.19}$$

Here N_i are the initial occupation numbers of each population and $\bar{N} = \nu/(\mu - \varepsilon)$. The new feature that appears in Equation (8.19) is the presence of oscillatory terms that reflect the cyclic nature of the system. It is not difficult to show that the auto- and cross-correlation functions characterising the populations also exhibit an oscillatory structure illustrated in Figure 8.5. In the case of three populations the period of the oscillation is commensurate with the decay imposed by the exponential factor so that the oscillations are barely visible. However, when more populations are included in the loop the decay time increases so that the oscillatory structure becomes more evident. This can be seen by noting that the mean of each of the populations governed by Equation (8.18) can be written more generally for the case $\mu_1 = \mu_2 = \mu$, $\varepsilon_1 = \varepsilon_2 = \varepsilon$, $\nu_1 = \nu_2 = \nu$ in the form

$$\bar{N}(t) = \bar{N} + \sum_{j=1}^{n} A_j \exp(a_j t), \tag{8.20}$$

where the exponents are given by

$$a_j = -\mu + \varepsilon\cos[2\pi(j-1)/n] + i\varepsilon\sin[2\pi(j-1)/n]; \quad j = 1,2,\ldots n.$$

Consider for example the time constant a_2:

$$a_2 = -\mu + \varepsilon\cos(2\pi/n) + i\varepsilon\sin(2\pi/n). \tag{8.21}$$

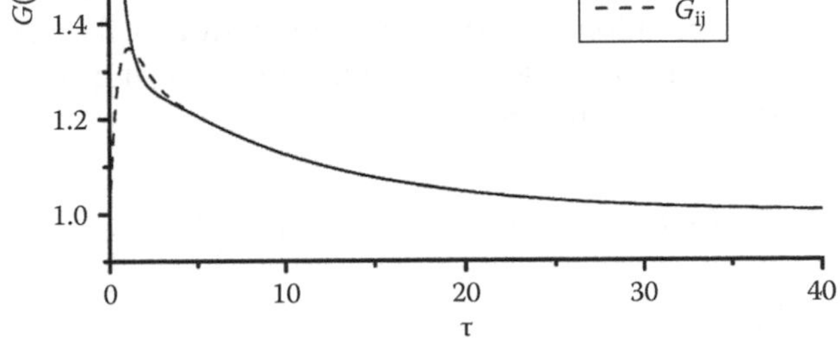

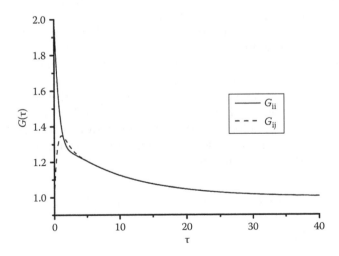

FIGURE 8.5
Auto- and cross-correlation functions for the three coupled-populations model. Parameters are $\nu = 0.5$, $\mu = 1$ and $\varepsilon = 0.9$.

Now let $\mu = \varepsilon + \quad /n$ and scale the immigration parameters with the number of populations $\varepsilon \to \varepsilon n$, $\nu \to \nu/n$, so that $\bar{N} = \nu/\quad$ is finite at long times. Substituting into Equation (8.21) we find that as the number of linked populations becomes large

$$a_2 \approx -\frac{\quad + 2\pi^2\varepsilon}{n} + 2\pi i\varepsilon. \tag{8.22}$$

Thus the solution (8.20) has at least one term of the form

$$A\exp[-(\quad + 2\pi^2\varepsilon)t/n]\cos(2\pi\varepsilon\, t) \tag{8.23}$$

that has a decay time much greater than its period of oscillation when n is large. This behaviour will also be manifest in the correlation functions characterising the stationary state and reflects the fact that as the number of linked populations with finite occupation number increases, the decay time of fluctuations increases but the characteristic frequency of oscillation in the approach to equilibrium remains the same.

8.5 Populations Coupled through Exchange Interactions

The previous sections of this chapter have been concerned with the evolution of a population of individuals of one species subject to immigration that is influenced by the presence of an *ensemble* of populations that may be of an entirely different type. Here we shall investigate the properties of a model in which there is *direct* exchange of individuals between just *two* populations of identical individuals. In the simplest model, individuals simply leave one population at a rate $\mu e N$ determined by the total number present in that population at the time of departure and immigrate into the second population at a rate that is independent of the number present in that population at the time. Assuming that the reverse process also takes place at a rate $\mu f M$ from the second population into the first, this requires the solution of the following rate equation for the joint distribution of finding N individuals in population 1 and M in population 2:

$$\begin{aligned}\frac{dP_{N,M}}{dt} &= \mu e[(N+1)P_{N+1,M-1} - NP_{N.M}] + \mu f[(M+1)P_{N-1,M+1} - MP_{N,M}] \\ &= F_{N,M}(t).\end{aligned} \tag{8.24}$$

Since no individuals are added to or leave the populations in this model, the total number of individuals is conserved and it is not difficult to show that in equilibrium the average number is the same as that present initially. This process would not be changed by internal monitoring, which will not perturb the internal population. However, external monitoring will effectively introduce a death rate so that the populations will be gradually depleted and eventually both populations will become extinct.

A more interesting model includes normal immigration and deaths in each population so that these will have non-zero mean occupation numbers whether or not exchange is taking place. In this case the appropriate rate equation is

$$\frac{dP_{N,M}}{dt} = \mu_1(N+1)P_{N+1,M} + \mu_2(M+1)P_{N,M+1} + \nu_1 P_{N-1,M} + \nu_2 P_{N,M-1} - (\mu_1 N + \mu_2 M + \nu_1 + \nu_2)P_{N,M} + F_{N,M}(t). \tag{8.25}$$

The evolution of the corresponding generating function is governed by the partial differential equation

$$\frac{\partial Q}{\partial t} = -[\mu_1 s_1 + \mu e(s_1 - s_2)]\frac{\partial Q}{\partial s_1} - [\mu_2 s_2 + \mu f(s_2 - s_1)]\frac{\partial Q}{\partial s_2} - (\nu_1 s_1 + \nu_2 s_2)Q \tag{8.26}$$

and the equilibrium solution of this is the product of two independent Poisson generating functions of the form

$$Q(s_1, s_2) = \exp[-(\bar{N}_1 s_1 + \bar{N}_2 s_2)], \tag{8.27}$$

where

$$\bar{N}_1 = \frac{\nu_1(\mu_2 + \mu f) + \nu_2 \mu f}{(\mu_1 + \mu e)(\mu_2 + \mu f) - ef\mu^2}, \quad \bar{N}_2 = \frac{\nu_2(\mu_1 + \mu e) + \nu_1 \mu e}{(\mu_1 + \mu e)(\mu_2 + \mu f) - ef\mu^2}.$$

These equilibrium mean values are independent of the initial states of the populations so that the memory preserving property of Equation (8.24) is lost by introducing separate immigration and death processes into the model. In the absence of the coupling, the population averages are given by $\bar{N}_{10} = \nu_1/\mu_1$ and $\bar{N}_{20} = \nu_2/\mu_2$ so that in equilibrium the average number of individuals in each population is changed by the presence of the exchange process:

$$\begin{aligned}\bar{N}_1 - \bar{N}_{10} &= -C\mu_2(\mu_2\nu_1 e - \mu_1\nu_2 f)\\ \bar{N}_2 - \bar{N}_{20} &= C\mu_1(\mu_2\nu_1 e - \mu_1\nu_2 f),\end{aligned} \tag{8.28}$$

which imply that the total number of individuals present is also changed,

$$\bar{N}_1 + \bar{N}_2 - (\bar{N}_{10} + \bar{N}_{20}) = C(\mu_1 - \mu_2)(\mu_2\nu_1 e - \mu_1\nu_2 f). \tag{8.29}$$

Here C is a positive constant that depends only on the system parameters. By setting $s_1 = s_2$ in Equation (8.27) we see that the sum $N_1 + N_2$ is also Poisson distributed. These results show that the difference between the coupled and uncoupled mean number of individuals is always of the opposite sign in the two populations, whereas the sign of the difference between the total number present in the coupled and uncoupled populations is determined by the particular values of the parameters of the model. Thus, although the exchange process itself conserves the total number of individuals in the two populations, the presence of additional deaths and immigrations can modify this quantity.

It is not difficult to obtain auto- and cross-correlation functions of the number of individuals in each population in the usual way from the first derivatives of Equation (8.26) with respect to s_1,s_2 evaluated at $s_1,s_2 = 0$, by averaging over the initial numbers present. The normalised quantities exhibit exponential decay on two distinct timescales dependent on the model parameters. These terms have amplitudes that are inversely proportional to the numbers of individuals present, which leads to the emergence of an interesting feature of the model when considering the measuring process. It can be shown that internal monitoring disrupts the statistics so that the measured populations are no longer Poisson distributed and results for the correlation properties are complicated in this case. External monitoring of individuals leaving the population provides a more tractable problem, not without some features of interest. In this case the generating function $q(s_1,s_2,z_1,z_2;T)$ characterising the number of events counted in a time interval T is obtained from the following modification of the rate Equation (8.26) as discussed in Chapter 3, Section 3.4.

$$\frac{\partial q}{\partial T} = -[\mu_1 s_1 + \mu e(s_1 - s_2) + \eta(s_1 - z_1)]\frac{\partial q}{\partial s_1}$$

$$-[\mu_2 s_2 + \mu f(s_2 - s_1) + \eta(s_2 - z_2)]\frac{\partial q}{\partial s_2} - (\nu_1 s_1 + \nu_2 s_2)q. \tag{8.30}$$

It has been assumed here that all individuals leave the mixed population by the same death process and are counted in the same way. Equation (8.30) has the solution

$$q(s_1,s_2,z_1,z_2;T) = \exp\left[-(\tilde{N}_1 s_1 + \tilde{N}_2 s_2)\right]\exp\left[-\eta T(\tilde{N}_1 z_1 + \tilde{N}_2 z_2)\right], \tag{8.31}$$

where $\tilde{N}_1, \tilde{N}_2$ are defined in Equation (8.27) but with $\mu_1 \to \mu_1 + \eta$, $\mu_2 \to \mu_2 + \eta$.

By setting $s_1 = s_2 = 0$ in Equation (8.31) we obtain the generating function for the joint probability of the number of individuals leaving each population in a time interval T. Like the internal population, this is seen to be jointly Poisson but with mean numbers $\eta\tilde{N}_1 T$ and $\eta\tilde{N}_2 T$. Moreover, it is readily shown that all the auto- and cross-correlation functions of the train of events associated with the individuals leaving the populations are unity. This is the same effect that was established in Chapter 3 for the train of events generated by individuals leaving a single death–immigration process, namely, the elimination of the time dependence of the correlation functions by the process of external monitoring. It is consequent on the monitoring process removing terms that are inversely proportional to the mean from the normalised correlation functions. Moreover, the inter-event times will be exponentially distributed with decay times depending solely on the mean number in each population. Therefore external monitoring will be unable to detect the transfer of individuals between the two populations and exchange is effectively a *hidden process* in this situation.

8.6 Summary

- We have investigated two kinds of coupling between evolving populations.
- In the first, multiple immigrations into each of two populations of potentially different kinds of individuals are controlled by the evolution of an ensemble of the other.
- We have shown that if the coupling constant and the death rate are equal and the same for both populations then these are geometrically distributed.
- In more general situations single immigrants and deaths have to be included in this model to avoid runaway populations or extinction. We showed that the normalised second factorial moment of the distributions then lies between one and two.
- We have shown that the power-law tail of an initially stable distribution is preserved by the process, though the equilibrium distribution is not stable.
- We have shown that cyclical coupling of populations leads them to exhibit an oscillatory approach to equilibrium, the number of periods in the decay time increasing with the number of linked populations.
- In the second kind of coupled population model, a death in one population generates an immigrant in another population of similar individuals and vice versa.
- We have shown that when simple immigration–death processes are linked by the exchange of individuals in this way, the internal populations remain Poisson processes and external monitoring cannot discern the existence of exchange, which therefore remains hidden.

Problems

8.1 Two death processes are coupled by multiple immigrations (Equation 8.1). Solve the two coupled time-dependent equations for the mean occupations of the populations assuming that at $t = 0$ the numbers present are N_0, M_0. Demonstrate that non-zero time-independent solutions exist only if $\mu_1\mu_2 = \varepsilon_1\varepsilon_2$. Show that if $\varepsilon_1 = \mu_2$ and $\varepsilon_2 = \mu_1$ then the total equilibrium mean is equal to the total occupation at the initial time.

8.2 Extend the results of Problem 8.1 by finding the differential equations that determine the evolution of the second moment of the populations. Show that when a time-independent solution exists, the equilibrium second factorial moments are given by

$$N^{[2]} = \frac{2\bar{N}\bar{M}}{3}\left(1+\frac{2\varepsilon_1}{\mu_1}\right), \quad M^{[2]} = \frac{2\bar{N}\bar{M}}{3}\left(1+\frac{2\varepsilon_2}{\mu_2}\right).$$

Show that in the special case $\varepsilon_j = \mu_j = \mu$ the equilibrium populations are characterised by geometric distributions.

8.3 The total number of individuals in a population is the sum of a large number M of sub-species, the number of each being governed by identical stable distributions. The number of sub-species varies with time according to an independent process. Find the generating function of the distribution of the total number of individuals in the population if the distribution of M is (1) Poisson and (2) geometric. Show that if $\bar{M}$ is increased without limit, whilst maintaining a non-zero value for the probability of finding no individuals present, the generating function reverts to the stable form in the case of Poisson number fluctuations whilst in the case of the geometric distribution it approaches $[1 + As^{\alpha}]^{-1}$.

8.4 A birth–death process, parameters λ, μ, is subject to multiple immigrations with probabilities that are controlled by a thermal (geometric) process with birth and death parameters $\sigma\lambda, \sigma\mu$ and a single immigration rate equal to its birth rate. Show that the total number of individuals in both populations is governed in equilibrium by a Laguerre distribution.

8.5 Individuals are exchanged between two populations subject to normal births, deaths and immigration, leaving one population at a rate proportional to the number of individuals present in that population and entering the other population as immigrants irrespective of the number of individuals already present. Assuming that the birth–death, immigration and exchange rates are the same for each population and that the death rate is greater than the birth rate: (a) find the equilibrium average and second factorial moment of the fluctuations in each population and also the correlation between the fluctuations, and (b) assuming that there are $N(0)$ individuals present in population 1 initially and $M(0)$ in population 2, find the average number present in the two populations at time t and the normalised auto- and cross-correlation functions characterising the populations in equilibrium. Prove that each population is characterised by two fluctuation times but that the fluctuation time of the sum of individuals in both populations is the same as when exchange is not present.

Further Reading

O.E. French, K.I. Hopcraft, E. Jakeman and T.J. Shepherd, 'Intrinsic and measured statistics of discrete stochastic populations,' Proceedings of the Royal Society A, **464**, 2929–2948 (2008).

E. Jakeman, K.I. Hopcraft and J.O. Matthews, 'Fluctuations in a coupled population model,' *Journal of Physics A: Mathematical and General*, **38**, 6447–6461 (2005).

9

Doubly Stochastic Processes

9.1 Introduction

In this chapter we study a simple way by which a positive-going continuous random variable can be associated with discrete events. In this model, events are triggered randomly at a rate that is governed by the magnitude of the variable. Thus we start by assuming that a 'variable' of constant magnitude will generate a proportionate constant rate of purely random and uncorrelated events. The number of events in a given time interval will then be Poisson distributed with a mean that is equal to the event rate multiplied by the counting time. When the variable changes sufficiently slowly with time it is assumed that the event rate follows suit, generating a Poisson series of events with a locally varying mean value. A model that combines a random Poisson series of events and a continuous random variable in this way is called a *doubly stochastic Poisson process* (Figure 9.1).

Since, by definition, the event rate generated by a doubly stochastic Poisson process has an additional variation imposed by the driving continuous random variable, its relative variance will always be greater than or equal to unity, that is, greater than that of a Poisson distribution having the same mean. This implies that the model cannot be used to characterise a train of sub-Poisson or anti-bunched events of the kind described in Chapter 5. Moreover, since the notion of even and odd has no meaning in the context of continuous processes, the doubly stochastic Poisson process cannot model discrete events that have properties that distinguish between such states. Despite these limitations the model has the advantage of simplicity and has proved to be useful in a number of practical situations. Originally conceived as a means of characterising the breakdown of looms in cotton mills, it has been used extensively in quantum optics for describing photoelectric emissions in the detection of so-called classical light fields and has found application in the modelling of point processes in many other fields including medicine and network theory.

One further refinement of the model is required to take account of situations where the underlying continuous process varies significantly over the

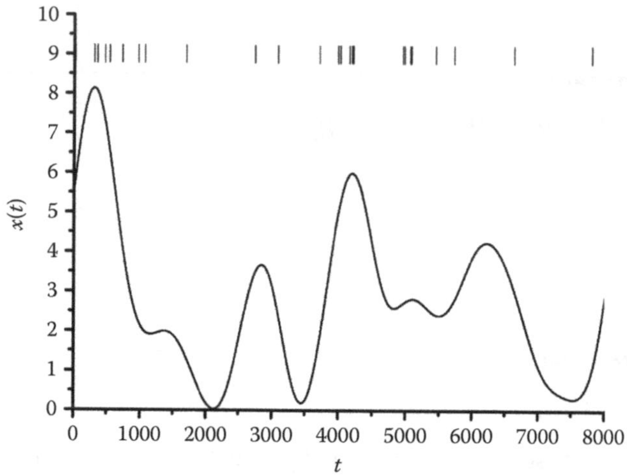

FIGURE 9.1
Doubly stochastic process. The upper series of random events have a time-varying mean, proportional to the random process x. See Chapter 11, Section 11.8 for details.

integration or counting time. In such cases the mean of the Poisson train of events is proportional to the *time average* of the continuous process over the interval in question. If $I(t)$ is the dimensionless magnitude or *intensity* of the continuous process, and we define the average or integrated intensity by

$$E(t;T) = \int_t^{t+T} I(t')dt', \tag{9.1}$$

where T is the integration or counting time, then we arrive at a fundamental formula for stationary ergodic doubly stochastic Poisson processes that relates the probability distribution of the number of events $p(n;T)$ in the interval T to the probability density $P(E)$ of the integrated continuous variable:

$$p(n;T) = \frac{1}{n!}\int_0^\infty (\xi E)^n \exp(-\xi E)P(E)dE. \tag{9.2}$$

Here, ξ is a constant of proportionality with dimensions of inverse time. In the monitoring of light by a photomultiplier detector, ξ represents the efficiency with which the photoelectrons (discrete photoevents) are produced in response to the intensity or *brightness* of light falling on the detector. Thus $0 < \xi \leq 1$ in this case and ξ is referred to as the *efficiency factor*.

Equation (9.2) can be generalised to relate the higher-order joint distributions of counts to the corresponding joint densities of the continuous variable generated by a stationary ergodic continuous stochastic process. For example, the joint distribution of finding n counts in an interval of duration T at time t and m in a second, *non-overlapping* interval of the same duration, at time t' is

$$p(n,m;T) = \frac{1}{n!m!}\int_0^\infty E^n\, dE \int_0^\infty E'^m\, dE' \exp(-\xi E - \xi E')P(E,E'), \tag{9.3}$$

where, referring to Equation (9.1), $E' = E(t';T)$.

9.2 General Statistical Relationships

In this section we derive some general relationships that are useful in the statistical analysis of doubly stochastic Poisson processes. We first recall the definition of the Laplace transform of a continuous function $l(t)$:

$$L(s) - \int_0^\infty dt\, l(t)\exp(-st). \tag{9.4}$$

Consider now the effect of multiplying both sides of Equation (9.2) by $(1 - s)^n$ and summing over n. On the left-hand side this gives the generating function $q(s;T)$ for the discrete distribution as described in previous chapters (see Chapter 2 and also Chapter 4, Section 4.6). On the right-hand side we find the expression

$$\int_0^\infty dEP(E)\exp(-s\xi E). \tag{9.5}$$

Comparison with Equation (9.4) indicates that this is just a Laplace transform of the distribution of integrated intensity. Thus, in the case of a doubly stochastic Poisson process, *the generating function for the discrete distribution is equal to the Laplace transform of the probability density of the continuous variable* apart from a scaling by the efficiency factor ξ:

$$q(s;T) \equiv L(\xi s;T). \tag{9.6}$$

Now the Laplace transform of the probability density of a continuous variable is just the moment generating function since from Equation (9.4) we have

$$\left(-\frac{\partial}{\partial s}\right)^r L(s)\big|_{s=0} = \int_0^\infty dt\, t^r l(t). \tag{9.7}$$

Thus for the doubly stochastic Poisson process the factorial moments of the distribution of events are proportional to the moments of the probability density of the underlying continuous variable

$$\langle n(n-1)(n-2)\ldots(n-r+1)\rangle = \xi^r \langle E^r \rangle.$$

After normalisation the efficiency factors cancel and we find that the normalised factorial moments of the discrete distribution and the normalised moments of the continuous distribution are identical

$$n^{[r]} = \frac{\langle n(n-1)(n-2)\ldots(n-r+1)\rangle}{\langle n\rangle^r} \equiv \frac{\langle E^r \rangle}{\langle E\rangle^r}. \tag{9.8}$$

This relationship greatly simplifies testing of the model against real data and the measurement of parameters governing the underlying stochastic process. However, there is no analogue of Equation (9.7) near the point $s = 1$ that would be necessary for a simple relationship to exist between the discrete distribution (determined through Equation 2.21, Chapter 2, for example) and the probability density of the continuous process. Inversion of the Poisson transform (9.2) can only be achieved by a full inverse Laplace transform of Equation (9.6).

The aforementioned approach can be extended to obtain a relationship between the generating functions of the joint distribution of the number of events in a fixed time interval and that of the joint density of the integrated or time averaged continuous variable. Using the definition (3.47) (see Chapter 3) it is not difficult to establish that the normalised autocorrelation function of the number of events in disjoint (i.e., non-overlapping) intervals is identical to the normalised autocorrelation function of the integrated continuous process:

$$g(\tau;T) = \frac{\langle n(0)n(\tau)\rangle}{\langle n(0)\rangle^2} = \frac{\langle EE'\rangle}{\langle E\rangle^2}. \tag{9.9}$$

We have assumed that the continuous process is stationary and ergodic so that on the right-hand side of this equation E and E' depend on the running time and the sampling or counting time as indicated in Equation (9.3) with $t' - t = \tau$. Integration over the time interval T is implicit in the definition of $n(t)$, that is, $n = n(t;T)$. This means that Equation (9.9) is not valid when $\tau < T$, which implies overlapping intervals.

9.3 Commonly Encountered Doubly Stochastic Poisson Processes

In this section we discuss the discrete probability distributions corresponding to probability densities that characterise some well-known continuous processes. The most trivial example is the case of a constant intensity when $E = IT = I_cT$ so that

$$P(E) = \delta(E - I_cT). \tag{9.10}$$

Substituting this model into Equation (9.2) gives the distribution

$$p(n) = \frac{(\xi T I_c)^n}{n!}\exp(-\xi T I_c). \tag{9.11}$$

This is a Poisson distribution with mean $\xi T I_c$ and will be obtained whatever the duration of the counting time T if the intensity is constant.

A Poisson distribution is also obtained *even if the intensity is varying* provided that T is sufficiently long; for if the fluctuation time of the intensity is much shorter than the counting time then the integral in Equation (9.1) will average out the fluctuations so that the integrated intensity will appear to be constant and Equation (9.10) will again apply but now $I_c = \langle I \rangle$. This is an important asymptotic result that holds at sufficiently long integration times for any continuous process characterised by a *finite* fluctuation time. It provides a useful test case for both theoretical calculations and experimental measurement. The exact effect of integration is most easily taken into account through result (9.6) that relates the moment generating function for the probability density of the continuous variable to the generating function corresponding to the discrete distribution.

A particularly simple model for the probability density of a fluctuating continuous variable is the negative exponential,

$$P(E) = \frac{\exp(-E/\langle E \rangle)}{\langle E \rangle}. \tag{9.12}$$

This distribution is often found to characterise, for example, the intensity fluctuations observed when laser light is scattered by random media. Substituting into Equation (9.2) gives the discrete distribution

$$p(n;T) = \frac{\langle \xi E \rangle^n}{(1 + \langle \xi E \rangle)^{n+1}}. \tag{9.13}$$

This is the familiar geometric or thermal distribution encountered first in Chapter 3. Note, however, that we have assumed here that the *integrated* intensity is exponential. In many practical applications, the appropriate stochastic process predicts that the intensity I is negative exponentially distributed, whilst E fluctuates rather less due to the averaging over T implicit in the definition (9.1). In this case result (9.13) will only apply for integration times that are much shorter than the characteristic fluctuation time of I and we must use Equation (9.6) for calculations in the more general case.

The gamma distribution provides a simple generalisation of the negative exponential model:

$$P(E) = \frac{b^{\alpha} E^{\alpha-1}}{\Gamma(\alpha)} \exp(-bE). \tag{9.14}$$

The constant b is related to the mean through $\langle E \rangle = \alpha/b$. This formula reduces to Equation (9.12) when $\alpha = 1$. Substituting into Equation (9.2) from Equation (9.14) gives the discrete distribution

$$p(n;T) = \begin{pmatrix} n+\alpha-1 \\ \alpha \end{pmatrix} \frac{(b/\xi)^n}{[1+(b/\xi)]^{n+\alpha}}. \tag{9.15}$$

This is the negative binomial distribution that we encountered as the equilibrium solution of the birth–death–immigration process in Chapter 4. Again, we must emphasise that the effect of integration is not taken into account in Equation (9.14). In practice, the gamma process predicts how a continuous variable I changes with time and we must use Equation (9.1) for calculations when the integration time is comparable to the characteristic fluctuation time τ_c of the variable. In fact an exact solution for the generating function of the probability density of an integrated gamma distributed variable is known. A discussion of the continuous gamma process is beyond the scope of this book but it is interesting to see the implications for the discrete distribution for this exactly solvable case. Thus it is found that

$$L(s;T) = \langle \exp(-Es) \rangle = \frac{\exp(\alpha\gamma)}{[\cosh y + (y/2\gamma + \gamma/2y)\sinh y]^{\alpha}}, \tag{9.16}$$

where $\gamma = T/2\tau_c;\, y^2 = \gamma^2 + 2\gamma s\langle E \rangle/\alpha$. This is structurally identical to the result (4.61b) obtained for the birth–death–immigration process (see Chapter 4). According to Equation (9.7) by calculating the derivatives at $s = 0$ this can be used to calculate both the moments of the integrated continuous variable and the factorial moments of the discrete distribution defined by Equation (9.2). Analytical expressions for the lower moments can be obtained without difficulty as we have seen in earlier chapters. Since the probabilities for small numbers of events are proportional to the derivatives of $q(s;T) \equiv L(\eta s;T)$ at $s = 1$, these can also be calculated fairly easily. However, the probability density of the continuous variable corresponding to Equation 9.16 is the inverse Laplace transform with respect to s and can only be evaluated using numerical techniques.

In fact the continuous gamma process may be interpreted as the high event-density limit of the externally measured birth–death–immigration process, and all the predictions of the latter model are identical with those for the event train generated by the doubly stochastic process using properties of the intensity predicted by the gamma process.

An important stochastic model arising in noise theory, the Rice process, relates to the vector sum of a constant (signal) component and a Gaussian distributed fluctuating (noise) component. If it is assumed that the angle between the two vectors is uniformly distributed over 2π radians, then fluctuations of the square of the amplitude (the intensity) of such a vector are governed by the probability density

$$P(E) = \frac{1}{\langle E_G \rangle} \exp\left(-\frac{E + E_C}{\langle E_G \rangle} \right) I_0 \left(\frac{2\sqrt{EE_C}}{\langle E_G \rangle} \right). \tag{9.17}$$

Here, I_0 is a modified Bessel function of the first kind, $E_C = I_c T$ is the 'integrated' intensity of the constant component, and $\langle E_G \rangle$ is the average integrated intensity of the Gaussian contribution. When this model is substituted into Equation (9.2) we obtain the discrete distribution

$$p(n;T) = \frac{\bar{n}_G^n}{(1+\bar{n}_G)^{n+1}} \exp\left[-\frac{\bar{n}_C}{(1+\bar{n}_G)} \right] L_n \left[-\frac{\bar{n}_C}{\bar{n}_G (1+\bar{n}_G)} \right]. \tag{9.18}$$

The constants in this formula are just η times the corresponding mean integrated intensities in Equation (9.17) and L_n are Laguerre polynomials. The distribution (9.18) is used as a model for the photon statistics of a mixture of laser light that has been scattered by a random medium with some of the original unscattered light. It is seen to be the special case $\alpha = 1$ of the class of Laguerre distributions that were introduced in Equation (2.11) (see Chapter 2). Indeed, the entire class of these discrete distributions can be generated through the doubly stochastic Poisson process from a generalisation of (9.17) based on the addition of a constant vector to one whose square modulus is gamma distributed. Again, we must emphasise that result (9.18) assumes that fluctuations in the *integrated* intensity are governed by the probability density (9.17). In practical applications it is usually the *instantaneous* intensity that has this distribution so that (9.18) will only be asymptotically correct if T is much less than the fluctuation time of the Gaussian variable.

The last example we consider in this section is a case where the continuous variable is itself a compound process. In particular if the mean of a negative exponential variable is modulated by a gamma variable we obtain the class of *K*-distributions (probability densities)

$$P(E) = \frac{2b}{\Gamma(\alpha)}(bE)^{\frac{\alpha-1}{2}} K_{\alpha-1}(2\sqrt{bE}). \tag{9.19}$$

Here α is the parameter in the gamma distribution (9.14), b is a constant that is inversely proportional to the mean value of E and $K_{\alpha-1}$ is a modified Bessel function of the second kind. The class of *K*-distributions (9.19) have found applications in many areas of science and technology, but in particular where the cross-section of scattered electromagnetic radiation is varying due to a changing density of scattering centres. Examples are microwave scattering from the sea surface, optical propagation through the atmosphere, and ultra-sound scattering from human tissue. Stochastic models predicting a variable I with this kind of probability density typically exhibit two distinct timescales: the shorter one associated with a negative exponential process and the longer one with the modulating gamma process used to model the scatterer density fluctuations. Thus the process of integration (9.1) creating the variable E can sometimes be tailored to average over the more rapid fluctuations so as to reveal properties of the scatterer density fluctuations. As in the previous models that we have considered (9.19) is usually only appropriate for very short integration times. The discrete distributions generated by the doubly stochastic Poisson process from the continuous densities (9.19) may be expressed in terms of Whittaker functions:

$$p(n;T) = \frac{n!}{\langle n\rangle^{\alpha/2}} \frac{\Gamma(n+\alpha)}{\Gamma(\alpha)} \exp\left(\frac{\alpha}{2\langle n\rangle}\right) W_{-(n+\frac{\alpha}{2}),\frac{(\alpha-1)}{2}}\left(\frac{\alpha}{\langle n\rangle}\right). \tag{9.20}$$

The mean number of counts in this formula is given by

$$\langle n\rangle = \eta\langle E\rangle = \eta T\alpha/b. \tag{9.21}$$

At this point a brief comment on the correlation properties associated with doubly stochastic Poisson processes is in order. In the case of many commonly encountered continuous stochastic processes, time evolution of a characteristic variable may be quantified by measurement of a single statistical quantity: its autocorrelation function. We have seen in Section 9.2 that the equivalent property of the discrete variable generated by the doubly stochastic process is simply proportional to this. For more complicated models involving multiple timescales, such as *K*-distributed noise, it is often necessary to measure higher-order statistical properties in order to elucidate the process. However, it is usually possible to generalise relationships such as (9.3) to higher-order products such that the normalised moments and correlation functions remain just proportional to the equivalent discrete statistics.

In the case of the continuous gamma process we have seen that the discrete equivalent appears to be just the birth–death–immigration process of Chapter 4 and that the results obtained for the normalised moments and correlation functions for these are identical. However, it is important to emphasise that more careful comparison of the results for the case of finite integration time reveals that equivalence lies only with the *externally monitored* birth–death–immigration process. This is an important result that, historically, enabled the doubly stochastic representation to provide a useful model for quantum detection of classical light fields. This involves the counting of individual photons as they leave the light source, that is, external monitoring.

9.4 Continuous and Discrete Stable Variables

In Chapter 7 we have introduced the concept of statistical stability and developed some related *discrete* random processes. These predict number distributions with the characteristic long inverse power-law tails that are found to characterise properties of many natural phenomena as well as measures used by financial institutions and markets. However, the concept of statistical stability is well known in the theory of *continuous* random processes and it is useful to quantify the relationship between these and the discrete models that are the main concern of the present book. This can be done with the help of a doubly stochastic formulation of the kind described in the current chapter and in this section we shall give a brief account of the relevant arguments.

The properties of a continuous stable variable are captured most succinctly by the corresponding characteristic function. This is defined as the exponential Fourier transform of the probability density function:

$$C(u) = \int_{-\infty}^{\infty} dx\, p(x) \exp(iux).$$

In the case of a so-called Lévy-stable variable this has the most general form

$$C(u) = \exp\left[-a|u|^{\nu}(1 - i\beta\,\mathrm{sgn}(u)\Phi(u,\nu))\right]$$
$$a > 0, \quad |\beta| \le 1, \quad 0 < \nu \le 2. \tag{9.22}$$

Here, a is a scaling constant and

$$\Phi(u,\nu) = \begin{cases} \tan(\pi\nu/2), & \nu \ne 1 \\ -(2/\pi \ln|u|), & \nu = 1. \end{cases}$$

In the special case $\nu = 2$ the distribution is Gaussian but its symmetry is more generally controlled by the parameter β. In the case $\beta = 0$, the distribution is symmetric and defined for all x. The *characteristic* function then superficially resembles the *generating* function (7.4) (see Chapter 7) for the stable discrete distributions. However, this is misleading because the index in the case of the generating function for the discrete stable distributions lies in the range $0 < \nu \leq 1$ and the corresponding distributions exist only for positive numbers of individuals and are manifestly not symmetric. An important feature of Equation (9.22) to note therefore is that when $|\beta| = 1$ and the index falls in the range $0 < \nu < 1$, the distributions are one-sided on the half-line defined through $sgn(x) = \beta$.

We can now establish a connection between continuous and discrete stable distributions using the Poisson transform (9.2). We have seen in Chapter 7 that the discrete stable distributions are defined through the generating functions

$$Q(s) = \sum_{n}^{\infty} P(n)(1-s)^n = \exp(-As^\nu), \quad 0 < \nu \leq 1. \tag{9.23}$$

Here A is the analogous scaling constant that can be identified with the mean of n when $\nu = 1$, corresponding to the Poisson distribution. A Poisson transform of the form (9.2) requires that the continuous random variable is defined for $x \geq 0$ and as we have seen the continuous one-sided stable distributions are those for which $\beta = 1$ and $0 < \nu < 1$. This range coincides with that for which the discrete stable distributions are defined. Suppose now that the distributions corresponding to the generating function (9.23) have a representation of the form (9.2). This means that

$$Q(s) = \sum_{n=0}^{\infty} \frac{(1-s)^n}{n!} \int_0^\infty dx\, x^n p(x) = \frac{1}{2\pi} \int_{-\infty}^{\infty} du \frac{C(u)}{s+iu}.$$

Substituting for $C(u)$ from Equation (9.22) and taking $\beta = 1$, so as to include only distributions that are non-zero in the positive half plane with $0 < \nu < 1$, obtains the result

$$Q(s) = \exp(-as^\nu \sec(\pi\nu/2)). \tag{9.24}$$

Comparison with Equation (9.23) shows that the one-sided Lévy-stable distributions and the discrete stable distributions form a Poisson transform pair with $A = a\sec(\pi\nu/2)$.

Figure 9.2 compares the discrete and continuous distributions for the case $\nu = 1/2$. We have already seen in Chapter 7 (Equation 7.15) that for this value of the index

$$p(n) = \frac{2}{n!\sqrt{\pi}} \left(\frac{1}{2}A\right)^{n+\frac{1}{2}} K_{n-\frac{1}{2}}(A),$$

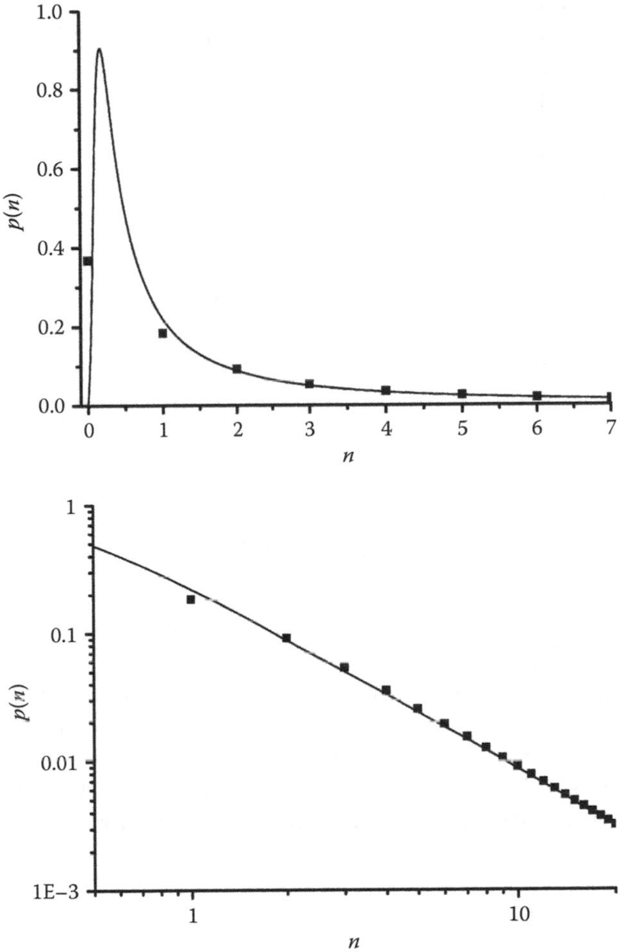

FIGURE 9.2
Comparison of the discrete and continuous stable distributions, with $\nu = 1/2$ and $a = A = 1$. The upper graph uses linear axes, and the lower graph shows the same functions but with logarithmic axes to reveal the power-law tails of the distributions.

whereas from Equation (9.22)

$$p(x) = \frac{a}{2x^{3/2}\sqrt{\pi}}\exp\left(-\frac{a^2}{4x}\right).$$

The discrete distribution is monotonically decreasing from a value of exp(−A) at $n = 0$. However, the continuous distribution is zero at the origin, then exhibits a peak near the origin followed by a long decline. The logarithmic plots shown in the lower graph of Figure 9.2 reveal the power-law tails of both distributions. It is interesting to compare the discrete stable distribution and its continuous counterpart when the index is close to unity (Figure 9.3).

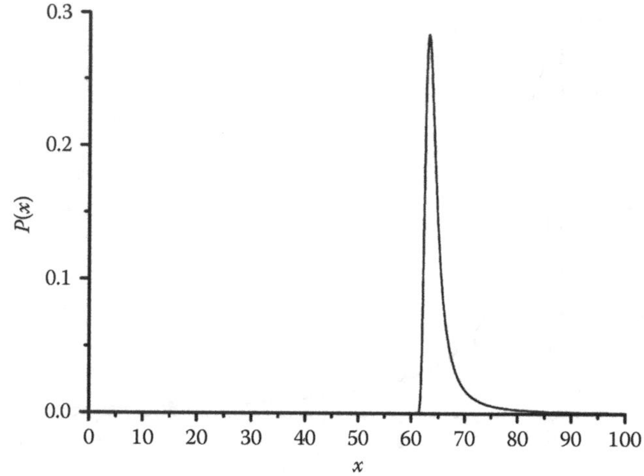

FIGURE 9.3
The continuous stable distribution for $\beta = 1$, $a = 1$ and $\nu = 0.99$.

The continuous distribution becomes comparatively highly peaked but still retains a power-law tail and so is not localised like the delta function it begins to resemble and finally attains when $\nu = 1$. On the other hand when ν is close to unity, for small values of n the discrete stable distribution is close to being Poisson distributed (Figure 9.4) and the power-law tail may not become manifest until n is very large. In Figure 9.4, $\nu = 0.99$ and deviation from a Poisson distribution only becomes apparent for values of n greater

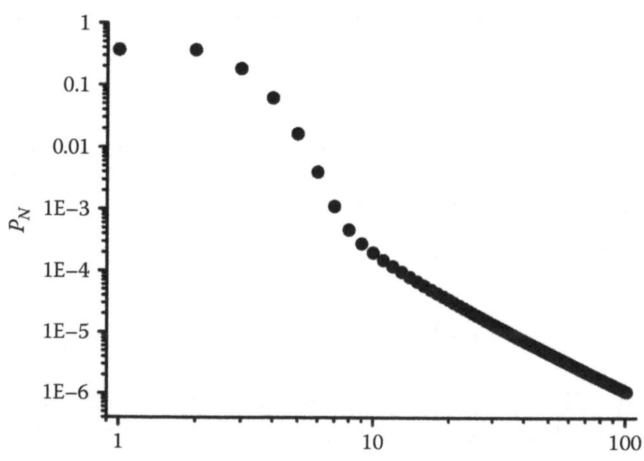

FIGURE 9.4
The discrete stable distribution for $A = 1$ and $\nu = 0.99$.

than 10. For numbers greater than this value, however, the existence of outlying rare events is apparent.

9.5 Doubly Stochastic Stable Processes

In the last section we were concerned with stable variables. However, it is possible to generalise the doubly stochastic model to include stable processes that are evolving with time. We have already encountered a discrete process of this kind in Chapter 7 where we studied the multiple immigration Markov process governed by the generating function equation

$$\frac{\partial Q}{\partial t} = -\mu s \frac{\partial Q}{\partial s} - a s^{\nu} Q. \tag{9.25}$$

In the case when $a = \nu\mu A$ this has Equation (9.23) as its stationary solution. Inverse Laplace transform of this equation leads, for a doubly stochastic Poisson process, to the integro-differential equation

$$\frac{\partial}{\partial t} p(x,t) = \mu \frac{\partial}{\partial x}[x p(x,t)] + \frac{A\mu\nu}{\Gamma(1-\nu)} \int_0^x \frac{dx'}{(x-x')^{\nu}} \frac{\partial p(x',t)}{\partial x'}. \tag{9.26}$$

This equation reveals the non-local and causality effects influencing the continuous process. Its solution is the conditional *continuous* density corresponding to the conditional *discrete* distribution that is the solution of Equation (9.25). These are related via the doubly stochastic representation. The joint distribution could be calculated from this solution in the usual way or directly from the joint generating function of the discrete process. However, if the latter approach is used then terms arising solely from the discrete nature of the variables must be eliminated. For example, the joint generating function for the stable multiple immigration process (9.25) was given in Chapter 7 as

$$Q(s,s';t) = \exp\{-A[s'^{\nu}(1-\theta^{\nu}) + (s+(1-s)s'\theta)^{\nu}]\} \tag{9.27}$$

where $\theta(t) = \exp(-\mu t)$. The continuous analogue of a joint Poisson process factorises into the product of two delta functions, however, and can only be obtained from the above result by suppressing the term in ss'. This is related to the suppression of terms in the distribution of order $\bar{n}^{-1}$ for according to the solution (9.24) as $\nu \to 1$, $A \to \bar{n} \to \infty$. Inspection of the joint statistics that we have derived earlier for other models suggests that terms like $ss'\theta$ are more generally associated with contributions to normalised autocorrelation functions of order $\bar{n}^{-1}$. The scaling that achieves suppression of terms like

$ss'\theta$ for all values of ν in Equation (9.27) is $s \to s(a/A)^{1/\nu}$, which, upon letting $A \to \infty$, gives

$$Q(s,s';t) = \exp\{-a[s'^{\nu}(1-\theta^{\nu}) + (s+s'\theta)^{\nu}]\}. \tag{9.28}$$

This result now represents the double Laplace transform of the one-sided stable process. In fact Equation (9.28) is a valid joint generating function corresponding to a stable single interval variable for any function $\theta(t)$ that decreases away from the origin, provided that $0 < \theta \le 1$. It can be Laplace inverted to obtain the joint stable density

$$p(x,x') = \frac{1}{[a^2(1-\theta^{\nu})]^{1/\nu}} p_{\nu}\left(\frac{x}{a^{1/\nu}}\right) p_{\nu}\left(\frac{x'-x\theta}{[a(1-\theta^{\nu})]^{1/\nu}}\right) H(x'-x\theta). \tag{9.29}$$

Here, the marginal one-sided Lévy density function $p_{\nu}(x)$ is the inverse Laplace transform of $\exp(-s^{\nu})$ and H is the Heaviside step function.

9.6 Summary

- We have shown how discrete events may be generated by a continuous variable through the notion of a doubly stochastic Poisson process.
- Statistical relationships between the continuous and discrete variables were derived, particularly the equivalence of the generating function of the distribution of the discrete variable and the Laplace transform of the continuous variable.
- We have given examples of continuous variables and their discrete counterparts that have been encountered in Chapter 7.
- The class of one-sided continuous stable variables is related to discrete stable variables.
- In the case of a doubly stochastic system, the joint probability density of a continuous stable process can be deduced by discarding terms in the generating function of the equivalent discrete distribution that are small in the large number limit.

Problems

9.1 Show that the generating function of the number distribution of events resulting from a doubly stochastic Poisson process is identical to the Laplace transform of the probability density of the continuous variable giving rise to it apart from a simple scaling of the mean. If a continuous variable E is uniformly distributed over the interval [0,u], what is the mean and variance of the corresponding discrete variable?

9.2 The continuous variable driving a doubly stochastic process is the modulus of the amplitude of a randomly phased sine wave, with the phase uniformly distributed between 0 and 2π. Show that the probability density of this variable is given by $P(E) = 2/\pi\sqrt{A_0^2 - E^2}$ in the region $0 \le E \le A_0$ and vanishes elsewhere. Show that the normalised second factorial moment of the associated discrete distribution is $\pi^2/8$ and that whilst $P(E)$ becomes asymptotically large for small A_0, the probability of finding no events approaches unity.

9.3 Suppose that n is the number of events generated by a doubly stochastic Poisson process. Express the average value of $(n + 1)^{-1}$ in terms of an average over a function of the driving continuous variable. Hence, evaluate $\langle 1/(n+1) \rangle$ for the continuous stable distribution of index 1/2.

9.4 A doubly stochastic Poisson process is driven by a continuous variable $E = x^2$ that is the square of the amplitude of a Gaussian process with joint density

$$p_2(x_1, x_2) = \frac{1}{2\pi\sigma^2(1-\rho^2)^{1/2}} \exp\left(-\frac{(x_1^2 + x_2^2 - 2x_1x_2\rho(\tau))}{2\sigma(1-\rho^2)} \right).$$

Show that the corresponding single-interval discrete probability distribution is negative binomial with index ½. What is the autocorrelation function of the discrete process? Assume that the efficiency factor is unity.

9.5 A train of impulses is well represented by a doubly stochastic discrete variable driven by a continuous variable that switches between E_0 and zero at random times. The proportion of time spent at zero is a. Write the probability density function of the continuous variable and hence calculate properties of the discrete variable. Suggest a method by which the value of E_0 could be determined from measurement of the statistics.

Further Reading

M. Bertolotti, 'Photon statistics,' in *Photon Correlation and Light Beating Spectroscopy*, H.Z. Cummins and E.R. Pike, eds., Plenum Press, New York, 1974.

S.B. Lowen and M.C. Teich, *Fractal-Based Point Processes*, chap. 4, Wiley, 2005.

M.C. Teich and P. Diament, 'Multiply stochastic representations for K distributions and their Poisson transforms', *J. Optical. Society. America.* A, **6**, 80-91, (1989).
K.D. Ward, R.J.A Tough, S. Watts, 'Sea Clutter: Scattering, the K-Distribution and Radar Performance', IET, (London), 2006.

10

Level Crossings of a Continuous Process

10.1 Introduction

The properties of zero or level crossings of continuous stochastic processes are of great importance to the topic of extremal statistics, which has numerous applications in biology, engineering, finance and the physical sciences. The zeros and level crossings provide one mechanism by which a discrete random process is formed from an underlying continuous variation. However the development that is required to treat these is different than what we have seen until now because they are not defined intrinsically but rather are a derived property of the continuous process. Nevertheless, we shall see that the models that we have developed can serve as useful tools with which to describe and explore the problem of crossings.

Why does such an apparently abstract problem as the crossings of a stochastic process have a wide currency in so many different application areas? One reason is that the 'problem' contains within it a variety of separate problems that translate to important practical questions such as how many large waves or gusts of wind are likely to hit an oil rig or skyscraper during its lifetime, or how long does one have to wait before such a large event occurs. Clearly knowing the answer to these types of questions enables tolerances and safety margins to be quantified.

The upper plot in Figure 10.1 shows part of a realisation of a continuous stochastic process $x(t)$. A useful technique for first isolating the zeros and then investigating their properties is to perform a simple but non-linear operation on the trace of $x(t)$ similar to the one that we encountered in Chapter 7, namely, 'clipping' that transforms the signal into a random 'telegraph wave' $T(t)$:

$$T(t)=\begin{cases}1 & x(t)\geq c\\ -1 & x(t)<c.\end{cases} \tag{10.1}$$

The locations of the level crossings coincide with those where the changes in parity of the telegraph wave occurs, indeed, the properties of the crossings define exclusively the properties of the telegraph wave. The telegraph wave

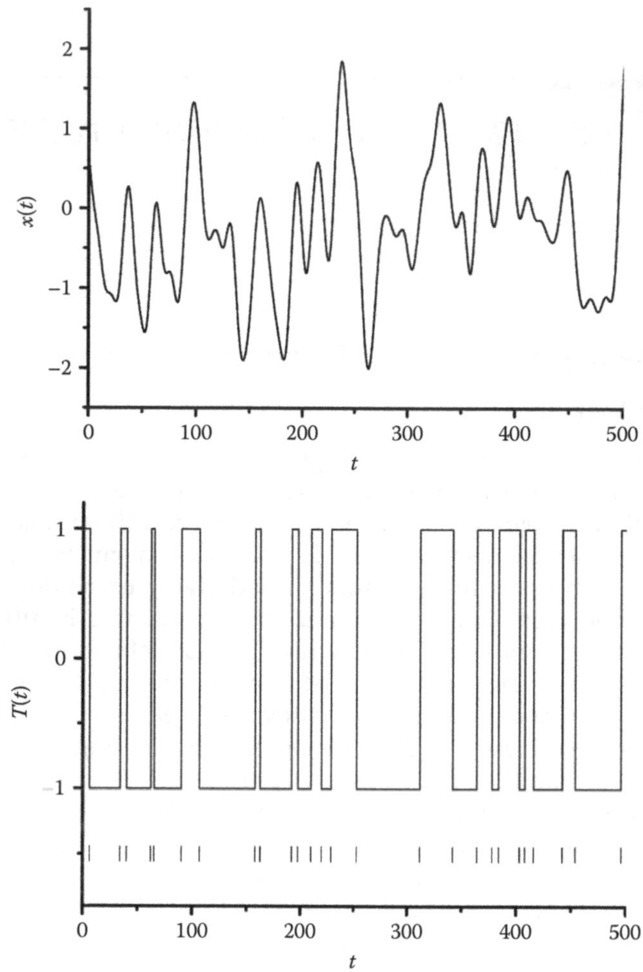

FIGURE 10.1
The first plot shows a section of a zero-mean Gaussian process with unit variance and a Gaussian autocorrelation function. The second plot is a telegraph wave formed from the zero-crossings of the process; below it is the series of events associated with the zero-crossings.

is defined by zero or level crossings of the process when $c = 0$ ('hard limiting') or $c \neq 0$ respectively. The lower plot in Figure 10.1 shows both the telegraph wave associated with the zero-crossings and the times at which the changes in parity occur. The location of the zeros forms another stochastic process that marks a discrete series of 'events.' We can pose questions about this new process, such as what are the moments of the crossings that occur in a given time interval and indeed what is their distribution? But we can also enquire about the relationship between the correlation properties of the crossings and those of the continuous process that generate them. These relationships prove to be surprisingly difficult to extract for the type of non-trivial continuous processes $x(t)$ that can occur in practical situations, and we will often have to resort to numerical techniques in order to quantify them. Embedded within the discrete process is another important continuous process that marks the length of the interval between two consecutive crossings and this quantity is frequently referred to as the 'interevent' or 'return' time.

The statistics of the telegraph wave are clearly defined by the probability density $p(x)$ for $x(t)$, for example,

$$\langle T\rangle = 1\times \mathrm{Prob}(x\geq c) - 1\times \mathrm{Prob}(x<c) = \int_c^{\infty} p(x')dx' - \int_{-\infty}^{c} p(x')dx'$$
$$= 1-2\wp(c)$$

where

$$\wp(x) = \int_{-\infty}^{x} p(x')dx'$$

is the cumulative distribution of x. Clearly, if the density function is symmetric, then the average value of the telegraph wave vanishes for zero-crossings.

An alternative process that has some advantages for theoretical developments is to rectify the telegraph signal, so that

$$R(t) = \frac{1}{2}(1+T(t)) = \begin{cases} 1 & x(t)\geq c \\ 0 & x(t)<c. \end{cases} \tag{10.2}$$

Part of this chapter will be concerned with the development of these numerical tools.

10.2 Generalisation of the Van Vleck Theorem

Forming a telegraph wave from an underlying continuous stochastic process is clearly a non-linear operation that affects the statistics and correlation properties of the clipped signal. The effect of hard limiting on the moments of the telegraph wave T is very simple, for if the continuous process has zero mean so too is the mean of T. The higher moments of T are either zero or unity depending upon whether odd or even moments are considered. The relationship between the autocorrelation functions of the continuous and clipped processes is more involved. Consider the correlation function of the rectified telegraph wave, $\langle R(0)R(\tau)\rangle = \langle R_1R_2\rangle$. The correlation product is either zero or unity, the latter value occurring when $x_1 = x(0)$ and $x_2 = x(\tau)$ are simultaneously above the threshold value, which we take to be zero. Hence, it follows that

$$\langle R_1 R_2 \rangle = \int_0^\infty dx_1 \int_0^\infty dx_2 \; p_2(x_1, x_2) \tag{10.3}$$

where $p_2(x_1,x_2)$ is the joint probability density function. The evaluation of the repeated integral in Equation (10.3) is generally problematical because of the paucity of useful or appropriate joint probability functions and the ability to perform the calculations. The most familiar process for which this is possible is the zero-mean Gaussian process with variance σ^2 and normalised autocorrelation function

$$\rho(\tau) = \langle x_1 x_2 \rangle / \sigma^2 .$$

Here we have used the abbreviation $\langle x(t)x(t+\tau) \rangle = \langle x_1 x_2 \rangle$. The corresponding joint distribution function is of the form

$$p_2(x_1, x_2) = \frac{1}{2\pi\sigma^2(1-\rho^2)^{1/2}} \exp\left(-\frac{\left(x_1^2 + x_2^2 - 2x_1 x_2 \rho(\tau)\right)}{2\sigma(1-\rho^2)} \right)$$

and the repeated integral in Equation (10.3) can be determined without any approximation to obtain

$$\langle R_1 R_2 \rangle = \frac{1}{4} + \frac{1}{2\pi} \arcsin(\rho(\tau)). \tag{10.4}$$

It follows from Equation (10.2) that for the un-rectified telegraph wave we have:

$$\langle T_1 T_2 \rangle = \frac{2}{\pi} \arcsin(\rho(\tau)). \tag{10.5}$$

This is known as the Van Vleck theorem or arcsine formula.

In the calculation of properties such as this, *characteristic functions* for the continuous processes often provide a useful mathematical tool. These are defined in terms of the Fourier transform of joint distributions with respect to their variables, for example,

$$C_2(\lambda, \mu) = \int_{-\infty}^{\infty} dx \int_{-\infty}^{\infty} dy p_2(x, y) \exp(i\lambda x + i\mu y).$$

Thus a generalisation of the Van Vleck theorem can be derived for joint *stable* statistics using the characteristic function

$$C_2(\lambda, \mu; r) = \exp\left(-\frac{A}{1+r^\nu} \left(|\lambda + r\mu|^\nu + |\mu + r\lambda|^\nu \right) \right). \tag{10.6}$$

Here $0 \le r(\tau) \le 1$ is a correlation coefficient or coherence function and $0 < \nu \le 2$. This model corresponds to a joint stable process that is positive or zero and although not unique, it is not difficult to show that it satisfies the correct limiting behaviour, namely,

$$C_2(\lambda,0;r) = C_1(\lambda); \quad C_2(0,\mu;r) = C_1(\mu),$$

$$C_2(\lambda,\mu;0) = C_1(\lambda)C_1(\mu); \quad C_2(\lambda,\mu;1) = C_1(\lambda+\mu).$$

Here C_1 is the characteristic function for a single interval stable probability density. In its simplest form this is similar to that discussed earlier in the context of stable *discrete* distributions, namely,

$$C_1(\lambda) = \exp(-A|\lambda|^{\nu}). \tag{10.7}$$

However, this model is valid specifically for a *symmetric* stable probability density of a *continuous* variable that can itself be positive or negative. Other two-sided stable distributions exist but are described by a generalisation of Equation (10.7). These will not be considered here and the interested reader is referred to the literature on continuous stable variables. Using the model (10.6) the telegraph wave autocorrelation function can be expressed in the form of an integral:

$$\langle T(t)T(t')\rangle \equiv \langle TT'\rangle = \frac{4r^{\nu}}{\pi^2}\int_0^{\infty}\frac{dp\, p^{\nu-1}}{(1+r^{\nu}p^{\nu})}\ln\left(\frac{1+p}{|1-p|}\right). \tag{10.8}$$

This is plotted against the coherence function r in Figure 10.2 for various values of the index. When $\nu = 2$ and $\rho = 1-(1-r)^2/(1+r^2)$, the model (10.6)

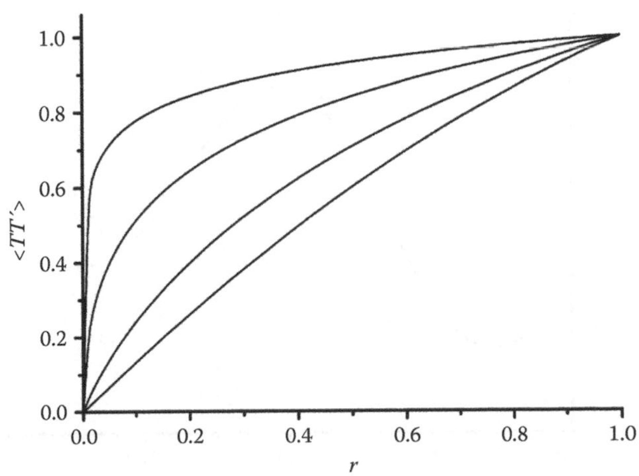

FIGURE 10.2
Telegraph wave autocorrelation function plotted against the coherence parameter r. From left to right, the curves are for $\nu = 0.2$, 0.5, 1.1 and 1.9.

reduces to the more familiar form for a Gaussian process and Equation (10.8) reduces to the arcsine formula (10.5).

10.3 Crossing Number Statistics of a Differentiable Process

Rice derived the average number of crossings for an arbitrary process heuristically but correctly and thereby stimulated an entire industry that has subsequently sought to generalise and make rigorous his approach. We shall outline the derivation in the same spirit as that adopted by Rice because this approach transparently reveals the underlying principle and is unencumbered by sophisticated mathematical machinery. However, we are safe in the knowledge that the validity of the result has been verified by exhaustive subsequent scrutiny. Figure 10.3 shows part of a realisation of a *continuous* and *differentiable* random process $x(t)$ in the vicinity of an upward zero-crossing, which occurs at time t that is located between the points t_0 and $t_0 + T$. In the vicinity of the zero-crossing we may approximate $x(t)$ by its Taylor series expansion about the point t_0,

$$x(t) \approx x(t_0) + \left.\frac{dx}{dt}\right|_{t=t_0} (t - t_0) = 0$$

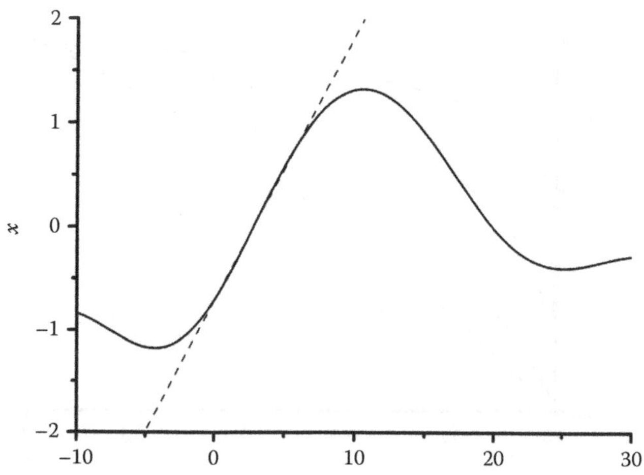

FIGURE 10.3
A random process in the vicinity of an upward zero-crossing.

and so the time at which the crossing occurs is

$$t = t_0 - \left.\frac{x}{\dot{x}}\right|_{t=t_0}$$

which clearly depends on both x and its derivative. Since we are considering an up-crossing, $x(t_0) < 0$ and $\dot{x}(t_0) > 0$ and clearly

$$t_0 < t_0 - \frac{x}{\dot{x}} < t_0 + T \tag{10.9}$$

from which we deduce that $-\dot{x}(t_0)T < x(t_0) < 0$. Now, the probability that the inequality (10.9) is satisfied is equivalent to the probability that a zero-crossing has taken place and can be evaluated by considering all the crossings that originate from all allowable negative starting points $x(t_0)$ with all possible positive values of $\dot{x}(t_0)$:

$$\int_0^{\infty} d\dot{x} \int_{-\dot{x}T}^{0} dx\, p(x,\dot{x},t_0) \approx T \int_0^{\infty} d\dot{x}\, \dot{x}\, p(0,\dot{x},t_0)$$

where $p(x,\dot{x},t)$ is the joint-probability density, which has been approximated by its value at $x = 0$. The same argument can be repeated for down-crossings, in which case the signs of $x(t_0)$ and $\dot{x}(t_0)$ are reversed, and the average number of zero-crossings occurring in an interval of length T becomes

$$\langle N \rangle = T \int_{-\infty}^{\infty} |\dot{x}|\, p(0,\dot{x},t)\, d\dot{x}. \tag{10.10}$$

This is the principle result that is valid for an arbitrary continuous, differentiable process, but to proceed further requires adopting a specific model for the joint density function. One that is ubiquitous and yields to further analysis is the stationary Gaussian process of zero mean, which, because x and $\dot{x}$ are statistically independent, has the form

$$p(x,\dot{x},\tau) = \frac{1}{2\pi\sigma\langle \dot{x}^2\rangle^{1/2}} \exp\left(-\frac{1}{2}\left(\frac{x^2}{\sigma^2} + \frac{\dot{x}^2}{\langle \dot{x}^2 \rangle}\right)\right).$$

In this formula, the variance of x and $\dot{x}$ are respectively σ^2 and $\langle \dot{x}^2 \rangle$. However, as we shall now see, these two quantities are related. Because the process is stationary, it follows that

$$\frac{d}{dt}\langle x(t)x(t+\tau)\rangle = 0.$$

Using this result and assuming further that the process $x(t)$ is continuous and differentiable it is straightforward to show by further differentiation that

$$\langle \dot{x}(t)\dot{x}(t+\tau)\rangle = -\frac{d^2}{d\tau^2}\langle x(t)x(t+\tau)\rangle = -\sigma^2\ddot{\rho}(\tau).$$

Consequently, $\langle \dot{x}^2\rangle = -\sigma^2\ddot{\rho}(0)$ and substituting into Equation (10.10) obtains the result

$$\langle N\rangle = \frac{T(-\ddot{\rho}(0))^{1/2}}{\pi}. \tag{10.11}$$

Note that this formula has meaning only when the second derivative of the autocorrelation function of the process exists in the form $1+\ddot{\rho}(0)\tau^2/2+\cdots$ near $\tau = 0$, which is in turn dependent on the process itself being continuous and once differentiable. Result (10.11) is easily generalised to determine the mean number of crossings of a level $x = u$, whereupon it is found that the mean is attenuated according to

$$\langle N\rangle = \frac{T(-\ddot{\rho}(0))^{1/2}}{\pi}\exp\left(-\frac{u^2}{2\sigma^2}\right). \tag{10.12}$$

Clearly, the higher the level, the fewer are the expected number of zero-crossings.

A formula for the second factorial moment of the number of zero-crossings of a differentiable Gaussian process has also been given and the interested reader is referred to the literature for a full derivation. The result obtained is

$$\langle N(N-1)\rangle =$$

$$= 2\left(\frac{T}{\pi}\right)^2 \int_0^1 dy(1-y)\frac{\sqrt{|A^2-B^2|}}{(1-\rho^2)^{3/2}}\left[1+\frac{B}{\sqrt{|A^2-B^2|}}\arctan\left(\frac{B}{\sqrt{|A^2-B^2|}}\right)\right]. \tag{10.13}$$

In the integrand of this expression

$$A = -\ddot{\rho}(0)[1-\rho^2(yT)] - \dot{\rho}^2(yT)$$
$$B = \ddot{\rho}(yT)[1-\rho^2(yT)] + \rho(yT)\dot{\rho}^2(yT).$$

Figure 10.4 shows a plot of (10.13) for the case of a Gaussian correlation function. It is not usually possible to evaluate the formula analytically, however, although it is possible to make predictions in some limiting

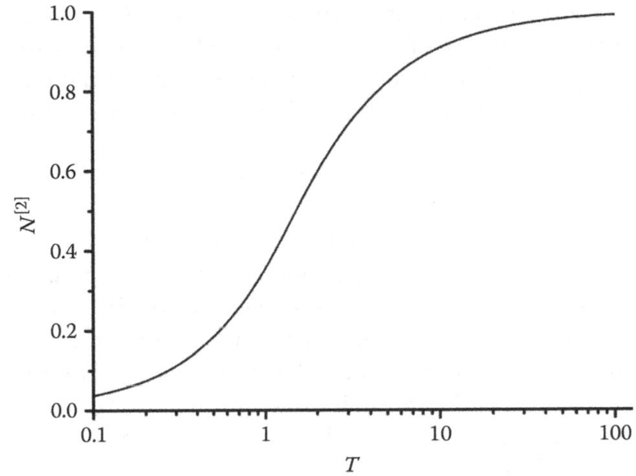

FIGURE 10.4
The normalised second factorial moment of the zero-crossings of a zero-mean Gaussian process in a time interval T. The autocorrelation function of the Gaussian process is $r(t) = \exp(-t^2)$.

situations. For example, in the small counting time limit, the correlation function

$$\rho(\tau) = 1 + \ddot{\rho}(0)\tau^2/2 + b|\tau|^{2+\mu}\cdots, \quad 0 < \mu < 2 \tag{10.14}$$

leads to the result

$$\langle N(N-1)\rangle = \left[\frac{\sqrt{4-\mu^2}}{\mu} + \arctan\left(\frac{\mu}{\sqrt{4-\mu^2}}\right)\right]\left(\frac{2b}{\pi^2\sqrt{-\ddot{\rho}(0)}}\right)T^{1+\mu} + O(T^{1+2\mu}). \tag{10.15}$$

This means that the *normalised* factorial moment scales as $T^{\mu-1}$, assuming the model expansion (10.14) for small T. In the special case $\mu = 1$ this quantity is independent of the interval length for small intervals. On the other hand, in the case of large intervals (10.13) predicts the asymptotic behaviour

$$\langle N(N-1)\rangle \sim 2\left(\frac{T}{\pi}\right)^2 \int_0^1 dy(1-y)|-\ddot{\rho}(0)| + O(T) \tag{10.16}$$

so that in this limit (C is a constant)

$$\langle N(N-1)\rangle \sim \frac{-\ddot{\rho}(0)}{\pi^2}T^2 + CT + \cdots = \langle N\rangle^2 + CT + \cdots$$

Therefore the normalised factorial moment approaches unity in this limit. However, the Fano factor exceeds unity, approaching the value $1 + \pi C/\sqrt{-\ddot{\rho}(0)}$. This behaviour has been encountered in Chapters 3, 4 and 6 confirms the

Fano factor as a more sensitive measure of deviations from Poisson statistics than the factorial moments of the number fluctuation distribution. This is illustrated in Figure 10.5, which shows a contour plot of the Fano factor as a function of the parameters μ and β assuming the model correlation function

$$\rho(\tau) = [1 + \beta(|\tau|/L)^{2+\mu}]\exp(-\tau^2/2L^2).$$

Result (10.10) can be recalculated for other processes for which the joint statistics are known, in particular processes derived from Gaussian such as the derivative of the chi-square process (the sum of the squares of several identical joint Gaussian processes). Of particular interest here, however, is a generalisation to the case of the joint stable model defined by the characteristic function (10.6). The equivalent joint distribution, $p_2(x,y)$, can be expressed in terms of the single interval stable density p_1 (corresponding to the characteristic function 10.7) in the form

$$p_2(x,y)dxdy = c(1+r^\nu)^{1/\nu} p_1(c(x-ry))p_1(c(y-rx))dxdy \tag{10.17}$$

where

$$c = \frac{(1+r^\nu)^{1/\nu}}{(1-r^2)}.$$

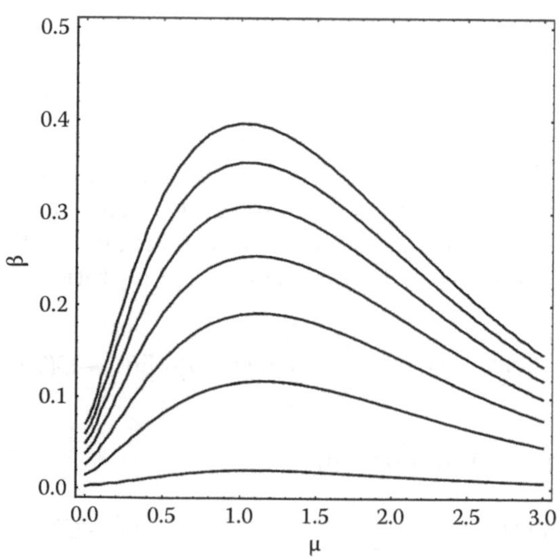

FIGURE 10.5
A contour plot of the Fano factor, F, for some different values of β and μ. The contours of equal F are spaced by 0.2, and the third contour from the bottom corresponds to $F = 1.0$.

We first note that the derivative $\dot{x}(t)$ is defined by

$$\dot{x}(t) = \lim_{\tau\to 0}\frac{x(t+\tau)-x(t)}{\tau}$$

so that if r can be expanded in the form $r = 1 - a\tau + \cdots$ for small τ then

$$p_2(x,\dot{x}) = \frac{2^{2/\nu - 1}}{a} p_1\left(2^{1/\nu - 1}\left(x + \dot{x}/a\right)\right) p_1\left(2^{1/\nu - 1}\left(x - \dot{x}/a\right)\right). \tag{10.18}$$

Substituting this result into the general formula (10.10) enables the mean number of crossings to be expressed after a little manipulation in the form

$$\langle N\rangle = \frac{4a\nu T}{\pi^2}\int_0^1 d\lambda \frac{\lambda^{\nu-1}}{(1+\lambda^\nu)^2}\ln\left|\frac{1+\lambda}{1-\lambda}\right|. \tag{10.19}$$

This must be evaluated numerically except for the special cases $\nu = 1$ and $\nu = 2$, which have the values $T/2$ and $\pi T/8$, respectively. Figure 10.6 shows a plot of (10.19) normalised against the Gaussian case, $\nu = 2$, as a function of ν. It can be seen that the crossing rate assuming Gaussian statistics is greatest. This is only to be expected since stable probability densities have long tails, implying that the process spends proportionately more time at high values, reducing the potential for zero-crossings.

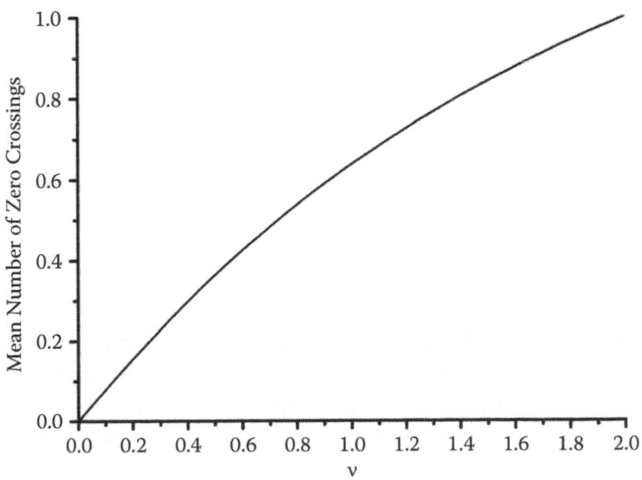

FIGURE 10.6
The mean number of zero-crossings for the joint stable distribution of Equation (10.6), normalised by the Gaussian case of $\nu = 2$.

10.4 Processes That Are Continuous but Not Differentiable

One of the most important assumptions we have made in the aforementioned calculations has been that of the *differentiability* of the process. Only when the derivative of the process exists is the basic formula (10.10) for the number of zero-crossings in the time interval T valid. Thus results (10.11) and (10.12) are finite only when the second derivative of the correlation function $\ddot{\rho}(\tau)$ exists at $\tau = 0$, which requires that near zero time delay $\rho(\tau) \sim 1 + \ddot{\rho}(0)\tau^2/2 + \cdots$. In the case of the stable model, calculations show that r may be related to a *notional* correlation coefficient $\tilde{\rho}$ through the formula

$$\tilde{\rho} = 1 - (1-r)^2 \big/ (1+r^{\nu})^{2/\nu}.$$

The assumption $r = 1 - a\tau + \cdots$ when τ is small is therefore seen to be equivalent to an expansion of this quantity in the form $\tilde{\rho}(\tau) \sim 1 - a^2\tau^2 / 2^{2/\nu} \cdots$ analogous to the Gaussian case. The important point here is that this is not consistent with Markovian behaviour of the kind that we have employed to formulate the discrete models considered earlier. Indeed, it is well known that a Gaussian-Markov process has a correlation function of the form $\exp(-\Gamma|\tau|)$ so that for small delay times it has an expansion of the form $\rho(\tau) \sim 1 - \Gamma|\tau| + \cdots$ and its derivative at the origin is discontinuous. This reflects the fact that the underlying process, though continuous, is not differentiable. A more general class of objects that display this kind of behaviour are the so-called fractal processes with correlation functions that exhibit power-law behaviour near zero delay time of the type

$$\rho(\tau) \sim 1 - \Gamma^{\alpha}|\tau|^{\alpha} + \cdots; \quad 0 < \alpha < 2. \tag{10.20}$$

In studying the crossings of such processes it is found that when their behaviour near any given level is examined, more crossings are revealed as the resolution of the inspection is improved. An important quantity for calculation in this case is the average of the process over a short time interval Δ:

$$X(t;\ ) = \frac{1}{\ } \int_{t}^{t+} dt' x(t'). \tag{10.21}$$

By definition $X(t)$ is differentiable. Moreover, in the case of joint Gaussian statistics it is also a Gaussian process and so we can calculate its zero-crossing statistics using the formulae of the last section. In particular the mean is given by Equation (10.11). However, the correlation function is now defined by

$$\langle X(0;\ )X(\tau;\ )\rangle = \frac{1}{\ ^2} \int_{0} dt \int_{\tau}^{\tau+} dt' \langle x(t)x(t')\rangle. \tag{10.22}$$

The second derivative of this quantity evaluated at zero may be written after normalisation

$$\ddot{\rho}_X(0;\;\;) = \frac{2}{\;^2}[\rho(\;\;) - \rho(0)].$$

When $\rho(\Delta)$ can be expressed, using Equation (10.14), as an even-powered expansion for small Δ, result (10.11) is obtained as expected. However, the fractal model (10.20) leads to

$$\ddot{\rho}_X(0;\;\;) = -2\Gamma^{\alpha}\;\;^{\alpha-2}. \tag{10.23}$$

Thus, as the resolution of the measurement is increased (Δ becomes smaller) the perceived number of crossings increases. This is consistent with an interpretation of the model correlation function (10.20) as imbuing a time trace of the Gaussian process with a cascade of finer and finer small-scale structures. It is important to draw a distinction here between (10.20) and models with the expansion (10.14). The latter 'sub-fractal' model characterises a process that is once differentiable, so that its level crossings are well defined, but whose derivative is not differentiable so that its local curvature is not well defined.

10.5 Distribution of Intervals between Zeros

Interevent time distributions can be derived from the distribution of zero or level crossings of a process, and the literature associated with these is longstanding and largely theoretical in nature but with few analytical results. A notable exception is the result obtained for a Gaussian process with the correlation function having the specific form

$$\rho(t) = \frac{3}{2}\exp\left(-|\tau|/3^{1/2}\right)\left(1 - \frac{1}{3}\exp\left(-|\tau|/3^{1/2}\right)\right).$$

This process has an interevent time density function expressible in terms of elliptic integrals with an exponential tail, a result that will not be presented here. Recent developments in this area have for the most part relied on numerical simulations. Figure 10.7 shows plots of probability distributions of the number of zero-crossings of a Gaussian process with the correlation function of the form

$$\rho(\tau) = \left[1 + b\left(\frac{|\tau|}{L}\right)^{2+\mu}\right]\exp\left(-\frac{|\tau|^2}{2L^2}\right). \tag{10.24}$$

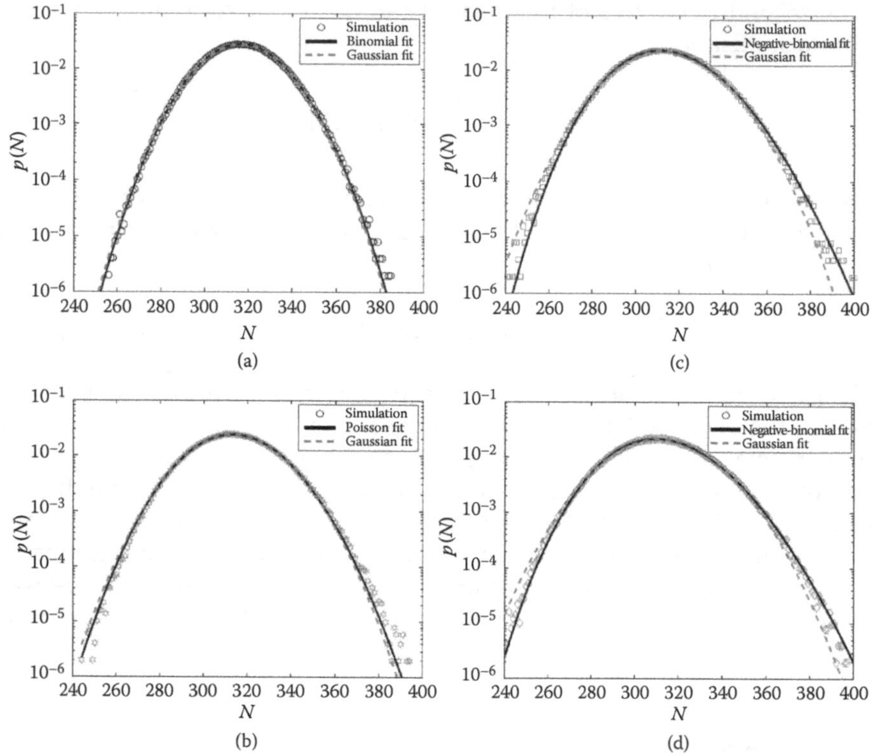

FIGURE 10.7
Probability distributions for the number of zero-crossings of a Gaussian process with autocorrelation given by Equation (10.24) occurring in an interval of length $T = 1000L$. Here, $\mu = 0.5$ and (a) $b = 0.051$, (b) $b = 0.14$, (c) $b = 0.167$ and (d) $b = 0.216$. The solid line shows a fit with (a) the binomial distribution, (b) the Poisson distribution and (c) and (d) the negative-binomial distribution. All results are with $L = 100$, and used 500,000 realisations. (From J.M. Smith, K.I. Hopcraft, and E. Jakeman, 'Fluctuations in the zeros of differentiable Gaussian processes,' *Physical Review E*, **77**, 031112, 2008.)

Comparison with three commonly encountered distributions indicates that when the Fano factor is less than unity the probability distribution is close to binomial, whereas when the Fano factor is large the negative binomial distribution provides a good fit. When the Fano factor is unity the distribution is closest to being Poisson distributed.

Figure 10.8 shows the corresponding correlation function of the number of fluctuations and also the distribution of intervals between crossings. The different behaviours displayed arise from interplay between the exponential factor and the power-law term in Equation (10.24). When the exponential term dominates (b sufficiently small) there is a smaller or inner scale separating the zero-crossings, the distribution is binomial-like and the Fano factor is less than unity. Under these conditions the number of crossings

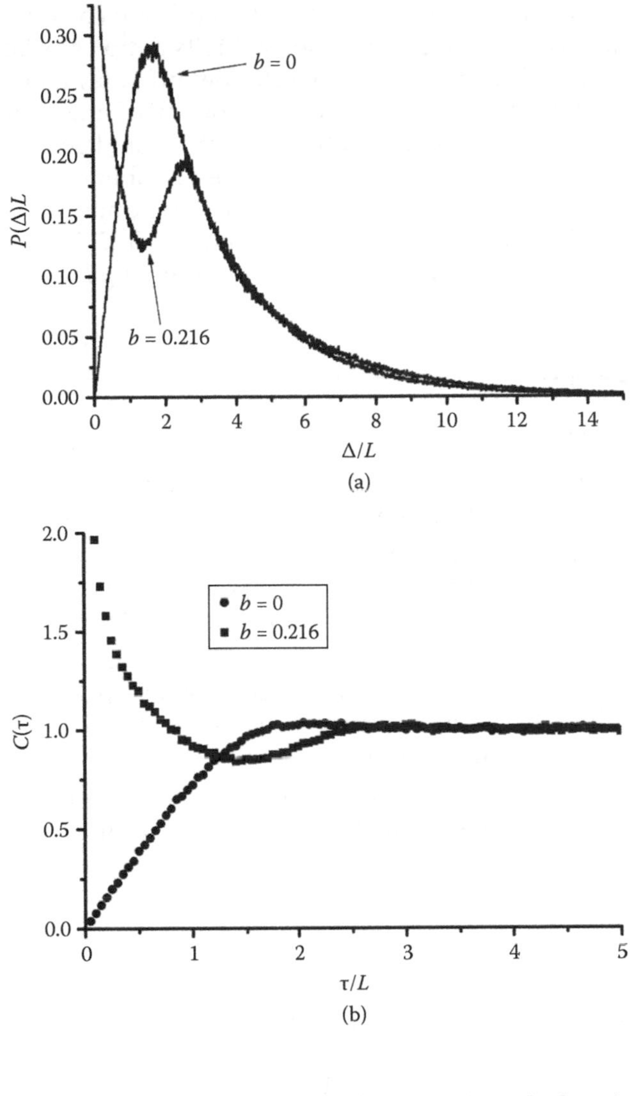

FIGURE 10.8
Simulation results for the number of fluctuations of zero-crossings of a Gaussian process with correlation function given by Equation (10.24). Here, $L = 100$ and $\mu = 0.5$. (a) The probability distribution of the interevent times, (b) the correlation function of the numbers counted in contiguous, nonoverlapping intervals of duration L/20. Note that the values for zero delay in (b) are off-scale (of order 60).

is anti-bunched as revealed by the autocorrelation function and the probability distribution of the time between crossings is unimodal with a peak near $\Delta \sim L$. As b increases clusters of crossings associated with the sub-fractal or power-law term in Equation (10.24) occur intermittently with cluster separation governed by the time constant of the exponential term. In this situation, the distribution is close to negative binomial, the Fano facto is greater than unity and the number of fluctuations are correlated, although the number autocorrelation function still exhibits a dip associated with the Gaussian factor in Equation (10.24), which now provides an outer scale or low frequency limit. The corresponding distribution of times between crossings becomes bimodal reflecting both the duration and separation time of clusters. In all cases the tail of the return time distribution eventually falls off exponentially as the number of crossings in long time intervals becomes approximately Poisson distributed. In this case result (2.57) (see Chapter 2) applies: $w_1(\ ) = r\exp(-r\ )$, where r is the mean crossing rate.

10.6 Summary

- We have shown how the level crossings of a continuous process can be used to define a telegraph wave or bimodal process.
- Relationships have been found between the autocorrelation functions of the original and derived processes for the case of Gaussian statistics and also for a joint stable statistical model.
- Results for the mean and second moment of the number of level crossings have been evaluated for differentiable processes and we showed how the case of processes that are continuous but not differentiable can be interpreted.
- We have discussed the problem of return time statistics and presented numerical results that indicated the relevant considerations in their interpretation.

Problems

10.1 If $R(t) = (1 + T(t))/2$ is a rectified telegraph wave, show that

$$\langle R^n \rangle = \frac{1}{2}(1 + \langle T \rangle)$$

for all values of n. If the continuous process is stationary so that $\langle T(t) \rangle = \langle T(t + t') \rangle$, then upon adopting the notation $T' = T(t')$, and so forth, show that the correlation functions

$$\langle RR' \rangle = \frac{1}{4}(1 + 2\langle T \rangle + \langle TT' \rangle)$$

$$\langle RR'R'' \rangle = \frac{1}{8}(1 + 3\langle T \rangle + \langle TT' \rangle + \langle TT'' \rangle + \langle T'T'' \rangle + \langle TT'T'' \rangle).$$

10.2 Carry through the calculations to obtain the Van Vleck theorem given by Equations (10.4) and (10.5) in the text.

10.3 Extend the argument in the text to show that the mean number of crossings of a level $x = u$ occurring in an interval T is given by

$$\langle N \rangle = T \int_{-\infty}^{\infty} d\dot{x}\, |\dot{x}|\, p(u, \dot{x}, t)$$

and evaluate this for a Gaussian process of zero mean, variance σ^2 and autocorrelation function $\rho(\tau)$ to show that the mean crossing rate is

$$\bar{r} = \frac{(-\ddot{\rho}(0))^{1/2}}{\pi} \exp\left(-\frac{u^2}{2\sigma^2}\right).$$

10.4 Evaluate the zero-crossing rate of the Gaussian processes having autocorrelation functions

$$\rho_1(\tau) = \exp\left(-\frac{\tau^2}{2L^2}\right)$$

and

$$\rho_2(\tau) = \left(1 + \frac{1}{\gamma}\frac{\tau^2}{L^2}\right)^{-\gamma/2}$$

where $\gamma > 0$. Deduce that any Gaussian process with autocorrelation function with expansion near the origin of the form

$$\rho(\tau) \approx 1 - \frac{\tau^2}{2L^2} + b\left(\frac{\tau}{L}\right)^{2+\mu} + \ldots$$

where $0 < \mu < 1$ will have identical zero-crossing rates. Contrast this with the zero-crossing rates associated with a fractal process for which the autocorrelation function has expansion

$$\rho(\tau) \approx 1 - b\frac{\tau^{2\mu}}{L^{2\mu}} + \ldots$$

where again, $0 < \mu < 1$.

10.5 An example of a strictly positive process is provided by the gamma process. This is a generalisation of the chi-square process, which can be formed by the sums of the squares of independent but statistically identical Gaussian processes. The gamma process has joint distribution

$$p(x,\dot{x}) = \frac{1}{2\Gamma(\alpha)\ {}^{\alpha+1/2}(-\pi\ddot{\rho}(0))^{1/2}} x^{\alpha-3/2} \exp\left(-\frac{1}{\ }\left(x + \frac{\dot{x}^2}{4\ddot{\rho}(0)x}\right)\right)$$

with $\alpha > 1/2$. Use this to show that the crossing rate at level $x = u$ is given by

$$\bar{r} = \frac{2}{\Gamma(\alpha)}\left(\frac{u}{\ }\right)^{\alpha-1/2}\left(\frac{-\ddot{\rho}(0)}{\pi}\right)^{1/2} \exp\left(-\frac{u}{\ }\right)$$

and hence show that the maximum rate of crossings occurs at a level $u = (\alpha - 1/2)\Delta$. Contrast and comment upon the rate that obtains as $u \to 0$ with that for a Gaussian process.

Further Reading

K.I. Hopcraft and E. Jakeman, 'On the joint statistics of stable random processes,' *Journal of Physics A: Mathematical and Theoretical*, **44**, 435101 (2011).

P. Lévy, *Theorie de l' Addition des Variables Aléatoires*, Gauthier-Villars, Paris, 1937.

D. Middleton, *An Introduction to Statistical Communication Theory*, chap. 9, IEEE Press, New York, 1996.

S.O. Rice, 'Mathematical Analysis of Random Noise', reproduced in *Noise and Stochastic Processes*, N. Wax (ed.), Dover, New York, 1954.

G. Samorodnitsky, and M.S. Taqqu, *Stable Non-Gaussian Random Processes: Stochastic Models with Infinite Variance*, New York, Chapman & Hall, 1994.

J.M. Smith, K.I. Hopcraft and E. Jakeman, 'Fluctuations in the zeros of differentiable Gaussian processes,' *Physical Review E*, **77**, 031112 (2008).

11

Simulation Techniques

11.1 Introduction

Numerical simulation provides a very useful tool for the investigation of population models. As such, it complements analytical calculations. Analysis can provide formulae for the statistics of the population, whereas simulation can give actual examples of populations and their evolutions. This provides additional insight into the models; numerical techniques have been employed for this purpose in the present book. In the language of statistics, we can say that the analytical approach gives information about the behaviour of *ensembles* of populations, whereas numerical calculations generate *realisations* of the populations. However, numerical analysis can also give *estimates* of ensemble properties, by averaging data from many realisations. The results are only estimates because only a finite number of realisations can be averaged, whereas a true ensemble average includes contributions from every possible state of the population; thus the estimates are subject to errors and a certain amount of insight is required to understand these errors. Numerical estimates of ensemble properties can be useful for checking the results of analytical calculations or for providing results in regimes where such calculations are intractable. Conversely, some models, such as the stable processes discussed in chapter 7, may present difficulties for numerical analysis, because some ensemble properties are determined by very rare, large events.

The numerical results given in this book were generated using the mathematical programming language *Mathematica*®, and the discussion in the present chapter will be supported by examples of *Mathematica* formulae. However, it is hoped that these are sufficiently simple that even the reader unfamiliar with *Mathematica* will be able to follow them and, if required, rewrite them in a programming language of their choice.

In most cases, the simulation algorithms are neither long, nor complex, and require no more than modest amounts of computer power.

11.2 Interevent Intervals

The models investigated in this book describe events, such as births, deaths or immigrations, which occur at a single instant in time. Consider the equation for the evolution of the death–immigration process (Equation 3.18, Chapter 3)

$$\frac{dP_N}{dt} = -\mu N P_N + \mu(N+1)P_{N+1} - \nu P_N + \nu P_{N-1}. \tag{11.1}$$

The rates μ and ν can be linked to the probabilities that deaths or immigrations, respectively, occur during an infinitesimal time period dt. Thus, νdt is the probability that a single immigration occurs during dt. Note that, owing to the infinitesimal nature of dt, the probability of more than one event occurring can be neglected. Similarly, the probability of a death occurring in a population of N individuals is $\mu N dt$ (the probability of any given individual dying is therefore μdt). Clearly, the probability of any event occurring is the sum of these probabilities, thus we can define a total event rate r by $rdt = \nu dt + \mu N dt$. The same argument can be extended to include births and other events in the total rate.

The most efficient way to simulate these processes is to focus on the time intervals between events. Suppose that an event occurs at time t. What is the probability that a time interval T will pass without a further event? The probability that a time dt will pass is $1 - rdt$. The probability that time $2dt$ will pass is given by the product of two identical terms, $(1 - rdt)^2$, and so on. Thus we can write,

$$P(T) \approx (1 - rdt)^{\frac{T}{dt}}. \tag{11.2}$$

Taking logarithms of both sides and the usual limit as $dt \to 0$ gives

$$\ln(P) = \frac{T}{dt}\ln(1 - rdt) \to -rT. \tag{11.3}$$

Therefore,

$$P(T) = \exp(-rT). \tag{11.4}$$

This is the cumulative distribution of the interevent intervals; the distribution of the intervals themselves is simply the absolute value of the derivative of Equation (11.4) with respect to T, see Chapter 2 Equation (2.57). The simulation method proceeds by randomly generating time intervals between events, where the durations of the intervals follow the correct probability distribution. This method is described in Section 11.3. Then, a further procedure is used to decide what kind of event has occurred and the population is changed accordingly. Finally, the value of r is updated in accordance with the new population.

11.3 Transformation of Random Variables

Many software packages allow the user to generate pseudo-random numbers. Typically, these will be real numbers, uniformly distributed in some interval, such as 0 to 1. These can be transformed to give random numbers having the required distribution for the inter event time intervals. In general, if a function f is used to transform a random number x to give a new number y, the probability densities of the random variables are related by

$$p_y(y)\left|\frac{df}{dx}\right| = p_x(x). \tag{11.5}$$

Integrating both sides of this equation gives a result in terms of the cumulative distribution functions of x and y

$$P_y(y) = P_x(x). \tag{11.6}$$

Solving for y in terms of x results in

$$y = f(x) = P_y^{-1}(P_x(x)) \tag{11.7}$$

P_y^{-1} being the inverse of the cumulative distribution function of y. In the case of the present interest, $P_x(x) = x$ is the cumulative distribution of the uniform variable x, and $y = T$, the interevent time. P_y^{-1} can be found from Equation (11.4), resulting in the following form for $f(x)$

$$T = -\frac{1}{r}\ln(x). \tag{11.8}$$

In *Mathematica*, the uniformly distributed random number is generated by the function RandomReal[ ], and the following is a user-defined function that will generate the required time intervals, using result (11.8).

```
deltaT[r_]: = -1/r*Log[RandomReal[]]
```

Note that this follows the convention that only built-in *Mathematica* functions start with a capital letter, user-defined functions starting with lowercase. The asterisk (*) denotes multiplication; a space can also be used for this purpose, but we will use the asterisk for maximum clarity.

11.4 Simulating the Death–Immigration Process

Having generated the random time interval, one needs to decide which of the possible events actually takes place at the end of the interval. Consider the example of the death–immigration process. Here, $r = \nu + \mu N$. Clearly, the relative probabilities of immigrations and births need to follow the ratios ν/r and $\mu N/r$. This can be achieved by generating a random number between 0 and 1 and selecting an immigration if the number is less than ν/r and a death otherwise. The following is an algorithm for simulating the death–immigration process. This was used to produce the results in Chapter 2, Figures 2.3 and 2.4.

```
n = 40; t = 0; (*Initial population and time*)
results = {{t,n}};
mu = 1;nimmig = 1; (*Death and immigration rates*)
While[t<200,
      death = mu*n;(*Total rate for death process*)
      rate = death+nimmig;(*Rate for all events*)
      t1 = deltaT[rate];(*Time interval before next event*)
      t = t + t1;
      rand = RandomReal[]*rate;
      Which[          (*Which event has occurred?*)
           rand< = death, n = n-1(*Death*),
           True, n = n + 1 (*Immigration*)
           ];
      results = Append[results,{t,n}];
      ]
```

The parentheses (* and *) enclose comments and are ignored when the algorithm runs. The initial population is set to be n = 40 at time t = 0. The previously defined function deltaT is used to generate time intervals between events. The algorithm goes around a loop until the time t exceeds 200. Pairs of values of time and population number are stored in the list called 'results.' This list contains all the information relating to the process. Of course, considerable subsequent processing may be required to extract information of interest: see Problems 11.2 and 11.3, for example, at the end of this chapter.

11.5 Extension to More than Two Events

All of the simulation results in Chapters 3 and 4 used modifications of the algorithm described in the previous section, with varying numbers of terms contributing to the variable called 'rate,' and corresponding numbers of cases to choose, depending on how many different events can occur. For example, the algorithm that produced the upper time series in Figure 4.7 (Chapter 4) used the following code:

```
death = mu*n; (*Death process*)
leav = eta*n; (*Leaving population*)
birth = lam*n; (*Birth process*)
rate = death+birth+nimmig+leav;
t1 = deltaT[rate]; (*Time interval before next event*)
t = t+t1;
rand = RandomReal[]*rate;
Which[           (*Which event has occurred?*)
     rand< = death,n = n-1, (Death*)
     (rand>death)&&(rand< = death+birth),n = n+1, (*Birth*)
     (rand>death+birth)&&(rand< = death+birth+leav),n = n-1;
     results1 = {res1,t} (*Leaving population*),
     True,n = n+1 (*Immigration*)
     ];
```

Here, there are four different events that can occur: deaths, births, immigrations and individuals leaving the population. The leavers are stored in a separate list: results1. Note that the && symbol, which is used when dividing the number segment between 0 and 1 to give the probabilities of the required events, performs a logical AND operation.

By comparing the rates here with the terms in Equation (4.1) (see Chapter 4), one can see that they are simply the coefficients of the corresponding terms in P_N on the right-hand side. One can see, then, how to treat other processes for which the differential equation for the probability is known. For example, to simulate the limited births that occur in Equation (5.2) (see Chapter 5) one would simply include a rate term equal to $\lambda(U - N)$.

11.6 Processes with Multiple Immigrations

The multiple-immigration processes introduced in Chapter 6 admit the possibility of an arbitrarily large number of immigrations. In order to decide how many immigrations occur, one can still use a random number representing a line segment, but now this line segment has an infinite number of subdivisions, see Figure 11.1. The problem then becomes one of identifying which division of the line segment a particular random number falls in. For the death–multiple immigration process of Figure 6.2 (Chapter 6), this is a straightforward task, because the multiple-immigration rates form a geometric series and the partial sums of this series can be written very easily. We have

$$\nu_r = a\zeta^r \tag{11.9}$$

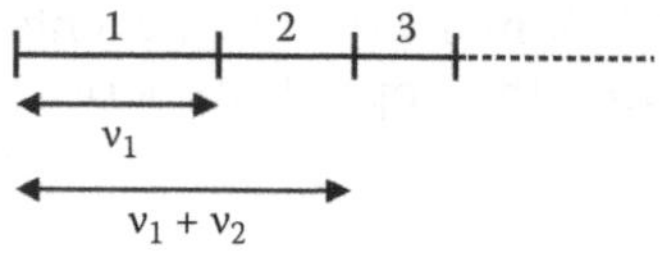

FIGURE 11.1
Division of a line segment in proportion to the probabilities of a given number of immigrants.

and the standard result for the partial sum of these terms is

$$x = \sum_{r=1}^{m} a\zeta^r = a\frac{\zeta(1-\zeta^m)}{1-\zeta}. \tag{11.10}$$

The simplest approach is to generate a random number between 0 and the value of (11.10) when m goes to infinity. Then, for a given value of this random number, x, one can calculate

$$m = \frac{\ln\left[\zeta + \frac{x}{a}(\zeta - 1)\right]}{\ln(\zeta)} - 1 \tag{11.11}$$

Here, Equation (11.10) has been re-arranged to give m in terms of x. Finally, m is rounded up to the nearest integer to find the number of multiple immigrants. The following is how this is implemented in the *Mathematica* code used to produce the lower plot in Figure 6.2 (see Chapter 6).

```
(* compute rates for the different processes*)
death = mu*n;(*Death process*)
rate = death+mig;
t1 = deltaT[rate];(*time interval before next event*)
rand = RandomReal[]*rate;
Which[
     rand<mig,j = Ceiling[Log[g+rand/a*(g-1)]/Log[g]-1];
     n = n+j;(*multiple immigration of order j*),
     (rand> = mig),n = n-1;(*death*)
     ];
     res = Append[res, {t, n}];
     t = t+t1;
```

Here, mig is the sum over all the immigration rates in Equation (11.10), that is, the limit of this expression when m goes to infinity.

A different death–multiple immigration process is encountered in Chapter 7 (see Figure 7.5). Here the partial sums of the immigration rates, Equation (7.11), cannot be so readily expressed in closed form. The approach taken here was to tabulate the partial sums to a sufficiently high order and to use the resulting table to find the order of multiple immigration for a given event. Clearly, this approach can become unwieldy for processes with very long-tailed distributions because very large numbers of multiple immigrations may be encountered; in Figure 7.5 the largest multiple immigration in the short time segment was only(!) of order 5500.

11.7 Gaussian Random Process

In Chapter 10 the level crossings of a continuous random process are investigated, in particular the Gaussian random process. Simulation of the Gaussian random process is greatly facilitated by the fact that the Gaussian is a *stable* distribution (as discussed in Chapter 7, Section 7.1) and thus a linear combination of Gaussian variables is also a Gaussian variable. A Gaussian distributed random number may be generated from uniformly distributed ones by the transform (11.7) using the inverse error function, but in practice it is more computationally efficient to use a scheme that does not involve special functions. A useful algorithm is based on the polar method. This starts with two random numbers, x_1 and x_2, that are uniformly distributed between 0 and 1, and transforms them to give two numbers, y_1 and y_2, that are the Cartesian components of a two-dimensional, zero mean, circular Gaussian distribution. These Cartesian components have unit variance. The transformation uses polar coordinates A and ϕ; x_1 is transformed to give a uniformly distributed phase angle ϕ (in radians)

$$\phi = 2\pi x_1. \tag{11.12}$$

Now the radial component A is given by $A^2 = y_1^2 + y_2^2$. It can be shown that the sum of the squares of two independent Gaussian variables has a negative exponential distribution. Thus x_2 is transformed into a negative exponentially distributed random variable, the square root of which gives the amplitude A. The cumulative distribution for the negative exponentially distributed random variable is given by

$$F_I(I) = 1 - \exp(-I/2). \tag{11.13}$$

Solving this for I gives the required transformation, via Equation (11.7),

$$A = \sqrt{I} = \sqrt{-2\ln(1 - x_2)}. \tag{11.14}$$

Using some simple geometry, y_1 is then obtained from

$$y_1 = A\cos(\phi) = \cos(2\pi x_1)\sqrt{-2\ln(1 - x_2)}. \tag{11.15}$$

y_2 is obtained similarly, using $\sin(\phi)$ instead of $\cos(\phi)$. y_1 and y_2 are zero mean uncorrelated Gaussian random numbers with unit variance. Other variances can be obtained by using suitable multiplying factors, and numbers with non-zero mean can be obtained by adding a constant. The complex number $z = y_1 + iy_2$ is a circular complex Gaussian variable. Here is a *Mathematica* function that generates Gaussian random numbers

```
nrand: = Cos[2.*Pi*RandomReal[]]*Sqrt[-
2.*Log[RandomReal[]+10.^-17]]
```

Note that $1 - x_2$ is just another random number between 0 and 1, allowing the expression to be slightly simplified, and a small positive number has been added inside the logarithm to avoid the problem of an infinite result if the round-off error causes the random number to be exactly 0.

A very useful method for simulating stationary Gaussian random processes is based on the discrete Fourier transform (DFT). The DFT is analogous to the ordinary Fourier transform but operates on discrete arrays of numbers rather than continuous functions. This method is sometimes referred to as filtering of Gaussian noise, because it is equivalent to starting with white (i.e., spectrally uniform) noise and imposing the required correlation via a filter (a multiplication in the frequency domain). The method can be implemented via a numerically efficient fast Fourier transform. Consider a one-dimensional array of N numbers $\{f_n\}$. The DFT of this is another array of length N, $\{F_m\}$, defined by

$$F_m = \frac{1}{\sqrt{N}} \sum_{n=0}^{N-1} f_n \exp\left(i \frac{2\pi mn}{N} \right). \tag{11.16}$$

The inverse transformation is

$$f_n = \frac{1}{\sqrt{N}} \sum_{m=0}^{N-1} F_m \exp\left(-i \frac{2\pi mn}{N} \right). \tag{11.17}$$

The similarity to the continuous Fourier transform (Equation 11.8) is evident. A continuous function $f(x)$ can be *sampled* at intervals Δx to produce a sequence $\{f_n\}$ from which a transformed sequence $\{F_m\}$ can be derived via Equation (11.16). Under ideal conditions this will be a good approximation to the Fourier transform of f, sampled at multiples of $\Delta k = 2\pi/(N\Delta x)$. There is a multiplying factor when going between the DFT and the continuous Fourier transform, which depends on the definitions used (a number of definitions are in common usage). When using Equations (11.16) and (11.17) to define the DFT and

$$\begin{aligned} F(k) &= \int_{-\infty}^{\infty} f(x) \exp(ikx)\, dx \\ f(x) &= \frac{1}{2\pi} \int_{-\infty}^{\infty} F(k) \exp(-ikx)\, dk \end{aligned} \tag{11.18}$$

for the continuous transform, the multiplying factor is $x\sqrt{N}$, that is, under ideal conditions the samples of F are given by the numbers F_m multiplied by $x\sqrt{N}$. However, significant information about the function f (and thus F also) will obviously be lost in this process if the function is non-zero outside the range of x values used in the sampling or if it varies too greatly within any Δx interval. For further discussion of these *sampling* issues the reader is referred to any standard textbook covering discrete Fourier transforms.

A Gaussian process can be completely described by its autocorrelation function. The random array representing a realisation of a Gaussian process is produced by the inverse DFT of the product of a set of weights, derived from the required autocorrelation function, and an array of uncorrelated complex Gaussian numbers with real and imaginary parts of zero mean and unit variance

$$z_n = \frac{1}{\sqrt{N}}\sum_{m=0}^{N-1} a_m w_m \exp\left(-i\frac{2\pi mn}{N}\right). \tag{11.19}$$

The complex Gaussian numbers a_m can be generated using results (11.12) to (11.15). Since z_n is a sum of Gaussians it is itself a complex Gaussian number. In fact, z_n is a set of samples of a circular Gaussian process, with real and imaginary parts that are uncorrelated and of equal variance.

Since the random numbers a_m are uncorrelated, that is, $\langle a_m a_q^* \rangle = \delta_{mq}$, the correlations of z are given by

$$\left\langle z_n z_p^* \right\rangle = \frac{1}{N}\sum_{m=0}^{N-1} w_m^2 \exp\left(-i\frac{2\pi m(n-p)}{N}\right). \tag{11.20}$$

By comparing this with Equation (11.17) it can be seen that the required weights $\{w_m\}$ are related to the autocorrelation through Equation (11.16):

$$w_m^2 = \sum_{r=0}^{N-1}\left\langle z_n z_{n-r}^* \right\rangle \exp\left(i\frac{2\pi mr}{N}\right). \tag{11.21}$$

Thus one can start with the required autocorrelation function and calculate its DFT, and the squares of the required weights are the DFT elements multiplied by $\sqrt{N}$. This is shown in the following *Mathematica* code:

```
n = Length[list1];
a = Table[nrand+I*nrand,{n}];
w = Sqrt[Chop[Fourier[list1]]*Sqrt[n]];
zlist = InverseFourier[a*w];
```

Here, `list1` is the autocorrelation function and `zlist` is the complex Gaussian time series. The function `Chop` removes any miniscule imaginary component of the Fourier transform, which may appear owing to the round-off error.

Note that in Equation (11.20) the difference $r = n - p$ can take on negative values, whereas in the original definition, Equation (11.17), positive values were assumed. Since the complex exponential is periodic, a negative value of r gives the same result as a value of $N - r$. This 'wrap-around' is a fundamental property of the DFT and appears in many guises. In the present case it is connected to a property of the realisations of the random process generated via the DFT, that of circular symmetry. The final point in the array of random values differs from the first by a factor of $\exp(2\pi im/N)$ in each term of the summation in Equation (11.19). However, exactly the same difference applies to each consecutive pair in the array; so the last and first values behave like any other consecutive pair and the array can be thought of as forming a continuous loop, its beginning following on from its end.

If a real Gaussian process is required, the real or imaginary part of z can be used. An example is given in Figure 11.2, in which a short segment of random process is generated. This starts with a Gaussian function for the normalised autocorrelation function, shown in the upper plot. Note that the zero delay point is the first point in the array and that the negative delays start from the end of the array, as discussed earlier. The real part of the generated realisation is plotted below the autocorrelation function. Note the way in which the end of the time series joins up with the beginning. If this circular correlation is not desired, then at least one correlation length must be dropped from the end or beginning of the time series.

11.8 The Doubly Stochastic Poisson Process

In the discrete random processes discussed so far, the rate at which events occur does not change until the population changes. This means that the simulations only need to calculate the time intervals between events. However, the doubly stochastic Poisson process introduced in Chapter 9 can have a rate that changes continuously. This means that a different simulation approach was required to produce Figure 9.1 (Chapter 9). The time axis is divided into short time intervals of length δt. The probability that there will be an event during this interval is taken to be $\gamma\delta t$. Here, γ is the time-varying rate, which is proportional to the value of the random variable $x(t)$. The simulation operates as follows: For each time step a random number is generated, which is uniformly distributed between 0 and 1. Only if this number is less than $\gamma\delta t$ does an event occur. Clearly, there are two requirements for the length of interval, δt. First, it must be sufficiently short that the continuous process that is

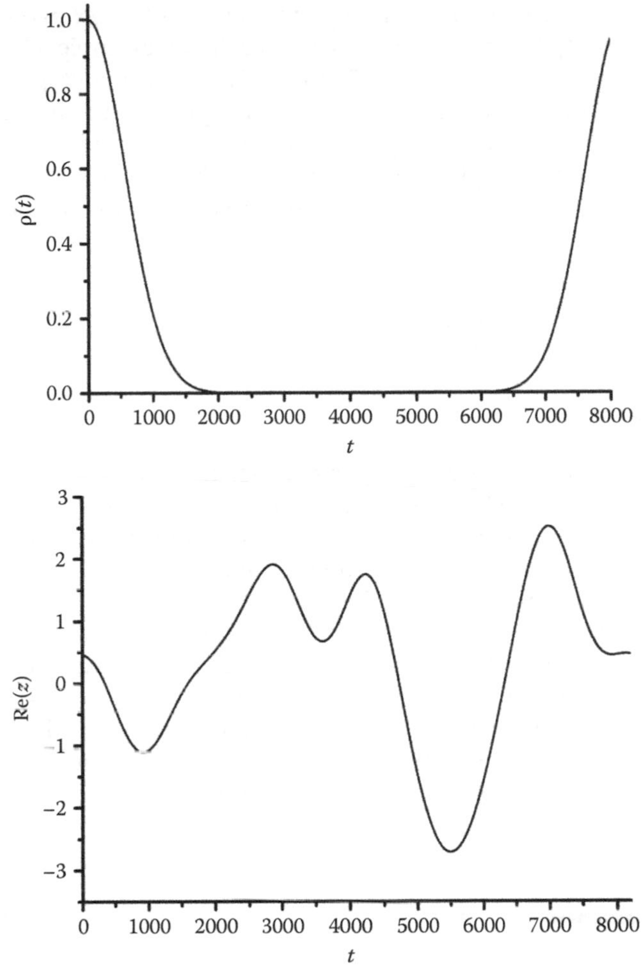

FIGURE 11.2
Upper plot is the required autocorrelation function of the random process, in this case a Gaussian function with a 1/e radius of 800. The generated process, z, is a circular complex Gaussian process and its real part, shown in the lower plot, is a real Gaussian process of unit variance. The imaginary part provides a second real Gaussian process.

determining the rate does not change significantly. Second, the probability $\gamma\delta t$ must be sufficiently small that the chances of two or more events occurring during one time interval are negligible. The probability of m events occurring is governed by the Poisson distribution of Equation (2.2) (Chapter 2)

$$P(m,\gamma\delta t) = \frac{(\gamma\delta t)^m}{m!}\exp(-\gamma\delta t). \tag{11.22}$$

Thus, the probability of more than one event is $1 - (1 + \gamma\delta t)\exp(-\gamma\delta t)$, which for small $\gamma\delta t$ is $\approx(\gamma\delta t)^2$. In Figure 9.1 (Chapter 9), for example, $\delta t = 1$ and $\gamma = 0.001x$. The largest value of x is 8.2, so the largest value of $\gamma\delta t$ is .0082. So there is less than one chance in 10^4 of a multiple event being missed in a given interval. Of course, the degree of accuracy required from a simulation depends on its actual application and needs to be decided accordingly. The *Mathematica* code that computes the events, given the time series $x(t)$, is

```
events = {};dt = 1.0;gam = 0.001;
Do[
  If[RandomReal[]<gam*x[[i]],
     events = Append[events,(i-1)*dt]
  ]
,{i,Length[x]}];
```

The result is a list of times at which events occur.

In the doubly stochastic Poisson process, the underlying continuous process needs to be positive. This could be achieved by taking the time series in Figure 11.2 and squaring it. However, the time series used in Figure 9.1 (Chapter 9) was generated in a slightly different way. It started with the autocorrelation function from Figure 9.1, but the process x was produced by summing the squares of both the real and imaginary parts of the generated complex Gaussian random process z. This kind of process, which has a negative exponential probability density, corresponds to the intensity of laser light scattered by a rough random surface, which is known as a speckle pattern, and plays an important role in light scattering.

11.9 Summary

- In this chapter we presented methods of simulating discrete population models. Examples have been given in the mathematical programming language *Mathematica*. We refer back to many of the figures in other chapters of this book for examples of the use of these algorithms.
- The key step is the generation of the random time intervals between events, where events may be births, deaths, immigrations and so forth. The resulting algorithms are relatively simple and compact, and do not require large amounts of computing power to run.
- We also discussed the generation of time series corresponding to continuous random processes and showed how the discrete Fourier transform can be used to generate a Gaussian random process with a given autocorrelation function.

Problems

11.1 By using the approach given in the algorithms in Sections 11.4 and 11.5, generate a time series of a population subject to births, deaths and immigrations using the same parameters as Figure 4.4.

11.2 Use the time series produced in Problem 11.1 to calculate the probability distribution for this process and compare with result (4.22) (see Chapter 4). Try using different lengths of time series and see how this affects the accuracy of your estimates of the probability. Note that the probability distribution is based on the *amount of time* for which the population has a certain number of individuals, not just the *number* of times it has that number of individuals, so you will need to use the time intervals between events as part of the calculation. Note, also, that to avoid bias you should discard a certain length from the beginning of the time series so that the starting value does not influence the result (see Equation 4.24 and the subsequent discussion).

11.3 Following on from Problem 11.1, use the time series to calculate the evolution of the mean of the BDI process (Equation 4.24, Chapter 4). Start each realisation of the process at $M = 10$, calculate a short time series and repeat many times to acquire enough data to compile an ensemble average. You will find it convenient to *resample* each time series so that the population number is given at regular intervals rather than just at the times at which events occur. This resampling algorithm will be required for investigating many of the time-dependent statistics discussed in this book.

11.4 Following the approach outlined in Section 11.5, write an algorithm to simulate the limited birth–death process described in Chapter 5. Include the process by which individuals leave the population and are counted (external monitoring). Investigate the distribution of the time intervals between leavers and compare with Equation (5.25).

11.5 Investigate the immigrant pair model of Chapter 5, Section 5.7, with internal monitoring. Note that the internal monitoring still needs to be treated in the same way as other events, even though it has no effect on the population number. Count the internal monitoring events using different time durations T. Calculate the corresponding Fano factors and verify that result (5.82) applies in the limit of large T.

11.6 Generate a time series of a Gaussian random process with Gaussian autocorrelation function. Turn this into a telegraph wave by applying the clipping function of Equation (10.1) (see Chapter 10). Thus, by averaging over a long time period or many realisations, verify the Van Vleck theorem (Equation 10.5). Note that autocorrelation functions are usually most efficiently calculated via discrete Fourier transforms. Mathematical programming languages will often have a built-in function that does this; *Mathematica* has a function `ListCorrelate`. Alternatively, methods of calculating autocorrelation functions can be found in signal processing textbooks.

Further Reading

E. Jakeman and K.D. Ridley, *Modeling fluctuations in scattered waves,* Taylor & Francis, New York, 2006.

D.E. Knuth, *The Art of Computer Programming: Volume 2, Seminumerical Algorithms,* 3rd ed., Addison Wesley Longman, 1998.

E. Renshaw, *Modelling Biological Populations in Space and Time,* Cambridge University Press, 1991.

H.J. Weaver, *Theory of Discrete and Continuous Fourier Analysis,* John Wiley & Sons, New York, 1989.

Glossary of Special Functions

G.1 Introduction

Throughout this book we have assumed that the reader is familiar with all the commonly used elementary transcendental functions. However, we have from time to time found it convenient to use some 'special' functions that arise naturally in the derivation of formulae and allow results to be written in a compact form. They are coded up in the mathematical software used for the computation of graphs and facilitate more complicated calculations that may involve the models described here as small sub-programmes.

It is not our purpose here to review the properties of special functions, but we aim to provide a little more information than would be expected in a conventional glossary. To this end we shall provide a basic definition for each function with detailed references to source texts. These will be listed in the 'References' section and provide the reader with a full treatment of the subject that can be studied off-line.

G.2 List of Special Functions in Alphabetical Order

Gamma (factorial) function $\Gamma(z)$

$$\Gamma(z) = \int_0^\infty dt\, t^{z-1} \exp(-t)$$

Note that if n is a positive integer $\Gamma(n) = (n - 1)!$
Ref. 1, Chapter 6; ref. 2, Chapter 8.1; ref. 3, Chapter 1

Hermite polynomials $H_n(x)$

$$H_n(x) = (-1)^n \exp(x^2)\frac{d^n}{dx^n}\left[\exp(-x^2)\right] = n!\sum_{m=0}^{[n/2]} \frac{(-1)^m (2x)^{n-2m}}{m!(n-2m)!}$$

Ref. 1, Chapter 22; ref. 2, Chapter 8.95; ref. 4, Chapter 10.13

Hypergeometric function $F(a,b;c;x) \equiv {}_2F_1(a,b;c;x)$

$$F(a,b;c;x) = \frac{\Gamma(c)}{\Gamma(a)\Gamma(b)} \sum_{n=0}^{\infty} \frac{\Gamma(a+n)\Gamma(b+n)}{\Gamma(c+n)} \frac{x^n}{n!}$$

$$= \frac{\Gamma(b)}{\Gamma(c)\Gamma(c-b)} \int_0^1 dt\, t^{b-1} \frac{(1-t)^{c-b-1}}{(1-tx)^a} \qquad c > b > 0$$

Ref. 1, Chapter 15; ref. 2, Chapter 9.1; ref. 4, Chapter 2

Incomplete gamma functions $\Gamma(a,x),\gamma(a,x)$

$$\Gamma(a,x) = \int_x^{\infty} dt\, t^{a-1} \exp(-t) \qquad \gamma(a,x) = \Gamma(a) - \Gamma(a,x) = \int_0^x dt\, t^{a-1} \exp(-t)$$

Ref. 1, Chapter 6; ref. 2, Chapter 8.35; ref. 4, Chapter 9

Laguerre polynomials $L_n^{\alpha}(x)$

$$L_n^{\alpha} = \frac{1}{n!} \exp(x)\, x^{-\alpha} \frac{d^n}{dx^n}\left[\exp(-x)\, x^{n+\alpha}\right] = \sum_{m=0}^{n} \frac{(-x)^m}{m!} \frac{\Gamma(n+\alpha+1)}{\Gamma(n-m+1)\Gamma(m+\alpha+1)}$$

Note that $H_{2n}(x) = (-1)^n 2^{2n} n! L_n^{-1/2}(x^2)$, $\quad H_{2n+1}(x) = (-1)^n 2^{2n+1} n! x L_n^{1/2}(x^2)$

Ref. 1, Chapter 22; ref. 2, Chapter 8.97; ref. 4, Chapter 10.12

Lerch transcendent $\Phi(x,s,v)$

$$\Phi(x,s,v) = \sum_{n=0}^{\infty} (v+n)^{-s} x^n \qquad |x| < 1, \quad v \neq 0, -1 \ldots$$

$$= \frac{1}{\Gamma(s)} \int_0^{\infty} dt\, \frac{t^{s-1} \exp(-vt)}{1 - x\exp(-t)} \qquad |x| < 1, \quad v > 0$$

Ref. 2, Chapter 9.55

Modified Bessel functions of the first kind $I_{\nu}(x)$

$$I_{\nu}(x) = \frac{(x/2)^{\nu}}{\Gamma(\nu + \frac{1}{2})\sqrt{\pi}} \int_{-1}^{1} dt\, (1-t^2)^{\nu - 1/2} \exp(-xt) \qquad \nu > -\tfrac{1}{2}$$

Ref. 1, Chapters 9 and 10; ref. 2, Chapter 8.4–8.5; ref. 4, Chapter 7

Modified Bessel functions of the second kind $K_\nu(z)$

$$K_\nu(x) = \frac{(x/2)^\nu \sqrt{\pi}}{\Gamma(\nu+\frac{1}{2})} \int_1^\infty dt\, (t^2-1)^{\nu-\frac{1}{2}} \exp(-xt) \qquad \nu > -\tfrac{1}{2}$$

Ref. 1, Chapters 9 and 10; ref. 2, Chapter 8.4–8.5; ref. 4, Chapter 7

$$K_{N+\frac{1}{2}}(x) = \sqrt{\frac{\pi}{2x}} \exp(-x) \sum_{k=0}^{N} \frac{(N+k)!}{k!(N-k)!(2x)^k}$$

Ref. 2, Chapter 8.46

Riemann zeta function $\zeta(x)$

$$\zeta(x) = \sum_{n=1}^{\infty} n^{-x} = \frac{1}{(1-2^{1-x})\Gamma(x)} \int_0^\infty dt\, \frac{t^{x-1}}{\exp(t)+1} \qquad x > 0$$

Ref. 1, Chapter 23.2; ref. 2, Chapter 9.51; ref. 4, Chapter 1.12

Whittaker function $W_{\lambda,\mu}(x)$

$$W_{\lambda,\mu}(x) = \frac{x^\lambda \exp(-\frac{1}{2}x)}{\Gamma(\mu-\lambda+\frac{1}{2})} \int_0^\infty dt\, \frac{(1+t/x)^{\mu+\lambda-\frac{1}{2}}}{t^{\lambda-\mu+\frac{1}{2}} \exp(t)}$$

Note that this can be expressed in terms of confluent hypergeometric functions. Ref. 1, Chapter 13; ref. 2, Chapter 9.22; ref. 4, Chapter 6.9

References

1. M. Abramowitz and I.A. Stegun, eds., *Handbook of Mathematical Functions*. Dover, New York, 1970.
2. I.S. Gradshteyn and I.M. Ryzhik, *Table of Integrals, Series and Products*, Academic Press, New York, 1980.
3. Bateman Manuscript Project, compilers, and A. Erdélyi, ed., *Higher Transcendental Functions, Volume 1*, McGraw-Hill, New York, 1953.
4. Bateman Manuscript Project, compilers, and A. Erdélyi, ed., *Higher Transcendental Functions, Volume 2*, McGraw-Hill, New York, 1953.

Index

Milton Keynes UK
Ingram Content Group UK Ltd.
UKHW040100071024
449327UK00019B/689